AS/A-LEVEL YEAR 1
STUDENT GUIDE

AQA

Chemistry

Inorganic and organic chemistry 1

Alyn G. McFarland

Nora Henry

PHILIP ALLAN FOR
HODDER
EDUCATION
AN HACHETTE UK COMPANY

Philip Allan, an imprint of Hodder Education, an Hachette UK company, Blenheim Court, George Street, Banbury, Oxfordshire OX16 5BH

Orders

Bookpoint Ltd, 130 Milton Park, Abingdon, Oxfordshire OX14 4SB

tel: 01235 827827

fax: 01235 400401

e-mail: education@bookpoint.co.uk

Lines are open 9.00 a.m.–5.00 p.m., Monday to Saturday, with a 24-hour message answering service. You can also order through the Hodder Education website: www.hoddereducation. co.uk

© Alyn G. McFarland and Nora Henry 2015

ISBN 978-1-4718-4369-3

First printed 2015

Impression number 5 4 3 2 1

Year 2019 2018 2017 2016 2015

This guide has been written specifically to support students preparing for the AQA AS and A-level Chemistry examinations. The content has been neither approved nor endorsed by AQA and remains the sole responsibility of the authors.

Cover photo: Ingo Bartussek/Fotolia

Typeset by Integra Software Services Pvt. Ltd, Pondicherry, India

Printed in Italy

Hachette UK's policy is to use papers that are natural, renewable and recyclable products and made from wood grown in sustainable forests. The logging and manufacturing processes are expected to conform to the environmental regulations of the country of origin.

Contents

▮ Getting the most from this book

Exam tips

Advice on key points in the text to help you learn and recall content, avoid pitfalls, and polish your exam technique in order to boost your grade.

Knowledge check

Rapid-fire questions throughout the Content Guidance section to check your understanding.

Knowledge check answers

1 Turn to the back of the book for the Knowledge check answers.

Summaries

- Each core topic is rounded off by a bullet-list summary for quick-check reference of what you need to know.

Exam-style questions

Commentary on the questions

Tips on what you need to do to gain full marks, indicated by the icon **e**

Sample student answers

Practise the questions, then look at the student answers that follow.

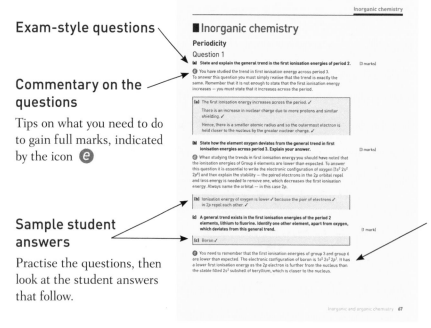

Commentary on sample student answers

Find out how many marks each answer would be awarded in the exam and then read the comments (preceded by the icon **e**), which show exactly how and where marks are gained or lost.

■About this book

This guide is the second of a series covering the AQA specifications for AS and A-level chemistry. It offers advice for the effective study of inorganic chemistry sections 3.2.1 to 3.2.3 and organic chemistry sections 3.3.1 to 3.3.6, which are examined on AS papers 1 and 2 and also as part of A-level papers 1, 2 and 3 as shown in the table below.

Section	Topic	AS Paper 1	AS Paper 2	A-level Paper 1	A-level Paper 2	A-level Paper 3
3.2 Inorganic chemistry						
3.2.1	Periodicity	✓		✓		✓
3.2.2	Group 2, the alkaline earth metals	✓		✓		✓
3.2.3	Group 7, the halogens	✓		✓		✓
3.3 Organic chemistry						
3.3.1	Introduction to organic chemistry		✓		✓	✓
3.3.2	Alkanes		✓		✓	✓
3.3.3	Halogenoalkanes		✓		✓	✓
3.3.4	Alkenes		✓		✓	✓
3.3.5	Alcohols		✓		✓	✓
3.3.6	Organic analysis		✓		✓	✓

Paper 1 of AS and A-level covers inorganic chemistry (periodicity, group 2 and group 7 from this book) and all physical chemistry sections in the student guide covering physical chemistry 1 in this series apart from kinetics (topic 3.1.5).

Paper 2 of AS and A-level covers organic chemistry (introduction to organic chemistry, alkanes, halogenoalkanes, alkenes, alcohols and organic analysis from this book) and all physical chemistry sections from the student guide covering physical chemistry 1 in this series apart from atomic structure (3.1.1) and oxidation, reduction and redox reactions (3.1.7). Paper 3 of A-level is synoptic and covers all topics and practical techniques.

This book has two sections:

■ The **Content Guidance** covers all of inorganic chemistry and organic chemistry for AS which are also part of A-level, and includes helpful tips on how to approach revision and improve exam technique. Do not skip over these tips as they provide important guidance. There are also knowledge check questions throughout this section, with answers at the end of the book. At the end of each section there is a summary of the key points covered. Many topics in these AS sections form the basis of synoptic questions in A-level papers. There are six required practicals at AS and notes to highlight these are indicated in the Content Guidance. These practicals and related techniques will also be examined in the A-level papers.

■ The **Questions & Answers** section gives sample examination questions on each topic, as well as worked answers and comments on the common pitfalls to avoid.

The Content Guidance and Questions & Answers section are divided into the topics listed on the AQA AS and A-level specifications.

Content Guidance

■ Periodicity

Periodicity refers to a repeating pattern of properties shown across different periods. Changes in melting point, atomic radius or first ionisation energy across the period are examples of periodic trends.

Classification

The elements are arranged in the periodic table
- by increasing atomic (proton) number
- in periods (horizontal rows)
- in groups (vertical columns)

The periodic table can be divided into blocks depending on which subshell the outer electron is found in. You need to be able to classify an element as an s, p, d or f block element. For example, sodium $1s^2\ 2s^2\ 2p^6\ 3s^1$ is an s block element. Its outer electrons are in an s subshell.

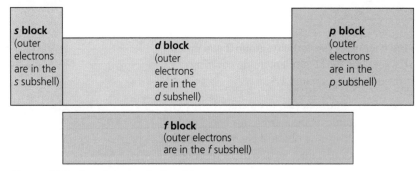

Figure 1 Blocks of the periodic table

Knowledge check 1

In which period and block is the element silicon found?

Physical properties of period 3 elements

Trends across period 3

1 Atomic radius decreases from Na to Ar. This is because:
   - there is an increase in nuclear charge across the period due to more protons
   - the shielding by inner electrons is similar, because the electrons that the elements gain across a period are added to the same shell
   - hence, there is a smaller atomic radius as the outermost electron is held closer to the nucleus by the greater nuclear charge

2 First ionisation energy shows a general increase from Na to Ar. This is because:
   - there is an increase in nuclear charge across the period due to more protons
   - the shielding by inner electrons is similar, as the electron is being removed from the same shell

- hence, there is a smaller atomic radius, as the outermost electron is held closer to the nucleus by the greater nuclear charge
- the first ionisation energies of elements in groups 2, 5 and 8 are higher than expected as a result of the repulsion due to the pairing of electrons in the first orbital

3 The melting point increases to Si and then decreases. The trend can be divided into three main sections:

a The metals. From Na to Mg to Al the metallic bond increases in strength, as there are more outer shell electrons that can be delocalised, giving a greater attraction between the electrons and the ions in the metallic structure.

b Silicon has a giant covalent structure and so has the highest melting point in the period, as a substantial amount of energy is required to break the large number of strong covalent bonds.

c The non-metals (phosphorus (P_4), sulfur (S_8) and chlorine (Cl_2)). These are non-polar, simple covalent molecules with low melting points. S_8 has the most electrons and so the greatest van der Waals forces of attraction between molecules. Argon is monatomic.

> **Exam tip**
>
> You must know these three trends for period 3 elements. However, you must also be able to apply them to other periods.

Summary

- Elements can be classified as *s, p, d* or *f* block elements. An *s* block element has its outer electrons in an *s* subshell. A *p* block element has its outer electrons in a *p* subshell.
- Across a period there is a decrease in atomic radius and an increase in first ionisation energy.
- The trend across the period in melting point can be explained in terms of structure and bonding of the elements.

■ Group 2, the alkaline earth metals

Trends from magnesium to barium

1 The atomic radius increases down the group.

The atoms of these elements all have an electronic configuration that ends in s^2:

Mg $1s^2\ 2s^2\ 2p^6\ 3s^2$

Ca $1s^2\ 2s^2\ 2p^6\ 3s^2\ 3p^6\ 4s^2$

Sr $1s^2\ 2s^2\ 2p^6\ 3s^2\ 3p^6\ 3d^{10}\ 4s^2\ 4p^6\ 5s^2$

Ba $1s^2\ 2s^2\ 2p^6\ 3s^2\ 3p^6\ 3d^{10}\ 4s^2\ 4p^6\ 4d^{10}\ 5s^2\ 5p^6\ 6s^2$

Down the group it can be noted that there is:

- an increasing number of shells, so there is more shielding
- hence, there is less attraction of the nucleus for the outer electrons, and so the atoms increase in size

2 The first ionisation energy decreases down the group. This is because:
   - there is an increase in atomic radius because there are more shells of electrons
   - there is more shielding of the outer electron from the nuclear charge, owing to increased number of shells
   - as a result, there is less nuclear attraction for the outer electron

3 The melting point of the elements decreases down the group.

The group 2 metals have metallic bonding — remember, this is the attraction between the metal cations and the delocalised electrons:
   - going down the group there are more shells of electrons and the metallic cations get bigger in size
   - as a result, the delocalised electrons are further away from the positive nucleus and the attraction between the positive cations and the delocalised electrons decreases
   - hence the metallic bonds are weaker and require less energy to break

Reactions of the elements with water

The reactivity of the group 2 metals increases down the group. This is evident in their reaction with water. Magnesium reacts slowly with cold water. The other group 2 elements (Ca, Sr and Ba) react readily with water.

The general equation is:

$$M + 2H_2O \rightarrow M(OH)_2 + H_2$$

For example:

$$Ca + 2H_2O \rightarrow Ca(OH)_2 + H_2$$

■ Observations for calcium, strontium and barium: heat is released; bubbles are produced; the metal disappears; a colourless solution is formed. The reaction increases in vigour as the group is descended.
■ When calcium reacts with water, the solution can appear milky (cloudy white) as calcium hydroxide is less soluble than either strontium hydroxide or barium hydroxide. Calcium sinks and then rises again owing to the rapid production of hydrogen, which raises the calcium granules.

Magnesium reacts readily with steam to produce an oxide and hydrogen:

$$Mg + H_2O(g) \rightarrow MgO + H_2$$

■ Observations: magnesium burns with a white light and a white powder is produced.

Solubility trends

You need to be able to recall the trend in solubilities of the group 2 hydroxides and sulfates in water. It is also important that you recall the solubilities of magnesium hydroxide and barium sulfate, as shown in Figure 2.

Solubility of the hydroxides

Magnesium hydroxide — virtually insoluble

↓ Increase in solubility

Barium hydroxide — soluble

Solubility of the sulfates

Magnesium sulfate — soluble

↓ Decrease in solubility

Barium sulfate — insoluble

Figure 2

Write the formula of the least soluble hydroxide of the group 2 metals from Mg to Ba.

> **Exam tip**
>
> The solubilities are used as the basis of identification tests. To distinguish between a solution of magnesium chloride and a solution of barium chloride add a few drops of sodium sulfate solution — a white precipitate of barium sulfate will form, as it is insoluble. Alternatively, add some sodium hydroxide solution and a white precipitate of magnesium hydroxide will form.

Uses of some group 2 compounds

- Barium sulfate: a barium sulfate suspension can be taken as a 'barium meal' to outline the stomach or intestines during X-rays. The barium sulfate is opaque to X-rays. Barium compounds are toxic, but barium sulfate is safe to use in this way, as it is insoluble.
- Magnesium hydroxide: magnesium hydroxide is used as an antacid in indigestion tablets to neutralise excess acid in the stomach and relieve indigestion.
- Calcium hydroxide: calcium hydroxide is used in agriculture to help reduce soil acidity.
- Calcium oxide or calcium carbonate: many of the flue gases from industrial processes contain acidic sulfur dioxide gas, which can lead to the production of acid rain. Calcium oxide and calcium carbonate are bases and can be used to remove the sulfur dioxide. The flue gases from, for example, the burning of the coal in power stations, are passed through a spray of finely ground calcium carbonate or calcium oxide suspended in water. The acidic sulfur dioxide reacts with the calcium oxide or carbonate and is neutralised:

$$CaCO_3 + SO_2 \rightarrow CaSO_3 + CO_2$$

$$CaO + SO_2 \rightarrow CaSO_3$$

Use of magnesium in the extraction of titanium

Titanium(IV) chloride can be reduced with a metal such as magnesium, at a temperature of around 700°C, to produce titanium metal.

$$TiCl_4 + 2Mg \rightarrow Ti + 2MgCl_2$$

The reaction must occur in an inert atmosphere of argon to prevent impurities that form in the presence of oxygen or nitrogen. These impurities make the metal brittle. The cost of production is high because magnesium is an expensive metal, a high temperature must be used, and the argon atmosphere also adds to the production costs.

Test for sulfate ion

Acidified barium chloride solution is added to a solution of the substance being tested. If a white precipitate is formed, then the substance must contain a sulfate. The precipitate is insoluble barium sulfate.

Ionic equation:

$$Ba^{2+}(aq) + SO_4^{2-}(aq) \rightarrow BaSO_4(s)$$

The barium chloride solution must be acidified with hydrochloric or nitric acid to remove any carbonate ions. Barium carbonate is a white insoluble solid that would be indistinguishable from barium sulfate.

> **Knowledge check 5**
>
> Explain why sulfuric acid should not be used to acidify barium chloride solution.

Summary

Going down group 2, the trends in properties are as follows:
- atomic radius increases
- first ionisation energy decreases
- melting point decreases
- reactivity increases
- solubility of the sulfates decreases
- solubility of the hydroxides increases

Some uses of group 2 metals include the following:
- Barium sulfate is used in a barium meal for X-ray patients. It is safe to use as it is insoluble.
- Calcium hydroxide is used to neutralise acidic soil. Magnesium hydroxide is used in indigestion remedies.
- Calcium oxide and calcium carbonate are used to neutralise sulfur dioxide in flue gases, which helps to prevent acid rain.
- Adding an acidified solution of barium chloride to a solution containing a sulfate ion produces a white precipitate of barium sulfate.
- Magnesium is used in the extraction of titanium from titanium(IV) chloride.

■ Group 7, the halogens

Trends in properties

Trends in boiling point and electronegativity

Halogen	Formula	Colour	State at room temperature	Trend in boiling point	Trend in electronegativity
Fluorine	F_2	Pale yellow	Gas	Increases	Decreases
Chlorine	Cl_2	Green	Gas		
Bromine	Br_2	Red-brown	Liquid	↓	↓
Iodine	I_2	Grey	Solid		

Table 1

1 Boiling points increase down the group.

   The M_r increases down the group, and there are more electrons, which means that there are increased induced dipole–dipole attractions (greater van der Waals forces) between the molecules. Hence, more energy must be supplied to break these stronger intermolecular forces.

2 Electronegativity decreases down the group.

   This is because the atomic radius increases, and the shielding increases, so the bonded electrons are further from the attractive power of the nucleus.

Trends in oxidising ability of the halogens

Oxidising agents are electron acceptors and are reduced in their reactions. The halogens are oxidising agents. The oxidising ability of the halogens decreases down the group.

Displacement reactions

The trend in oxidising ability of the halogens is illustrated by displacement reactions. A displacement reaction is one where the more reactive (strongest oxidising agent) halide will displace a less reactive one from a solution of its halide ions. The displacement reactions of the halogens are shown in Table 2.

	Chloride ion solution, e.g. NaCl(aq)	Bromide ion solution, e.g. NaBr(aq)	Iodide ion solution, e.g. NaI(aq)
Chlorine water (Cl_2)	No reaction	Chlorine displaces bromine from solution: $Cl_2 + 2NaBr \rightarrow 2NaCl + Br_2$ Ionic equation: $Cl_2 + 2Br^- \rightarrow 2Cl^- + Br_2$ Observation: colourless solution (NaBr) changes to orange solution (Br_2)	Chlorine displaces iodine from solution: $Cl_2 + 2NaI \rightarrow 2NaCl + I_2$ Ionic equation: $Cl_2 + 2I^- \rightarrow 2Cl^- + I_2$ Observation: colourless solution (NaI) changes to brown solution (I_2)
Bromine water (Br_2)	No reaction	No reaction	Bromine displaces iodine from solution: $Br_2 + 2NaI \rightarrow 2NaBr + I_2$ Ionic equation: $Br_2 + 2I^- \rightarrow 2Br^- + I_2$ Observation: colourless solution (NaI) changes to brown solution (I_2)
Iodine solution (I_2)	No reaction	No reaction	No reaction

Table 2

> **Exam tip**
>
> Oxidation numbers can be used to explain why displacement reactions are redox reactions. In $Cl_2 + 2Br^- \rightarrow 2Cl^- + Br_2$ the oxidation number of Cl decreases from 0 to –1 and it is reduced. The oxidation number of Br increases from –1 to 0 and it is oxidised. Both oxidation and reduction have occurred, so it is redox.

> **Knowledge check 6**
>
> Name the halogen that is the strongest oxidising agent.

> **Exam tip**
>
> The reactions of fluorine are not examined experimentally as fluorine is too dangerous to be used in the laboratory. The reactions of fluorine would follow the pattern for the other halogens.

Worked example

a Write the simplest ionic equation for the reaction of chlorine with potassium bromide solution.

b Write half-equations for the oxidation and reduction reactions that are occurring.

Answer

a The equation for this reaction is:

$Cl_2 + 2KBr \rightarrow 2KCl + Br_2$

The K^+ ion is a spectator ion, so is not included in the ionic equation.

The simplest ionic equation is:

$Cl_2 + 2Br^- \rightarrow 2Cl^- + Br_2$

b The half-equation for the conversion of chlorine molecules into chloride ions is:

$Cl_2 + 2e^- \rightarrow 2Cl^-$

This is a reduction half-equation as chlorine is gaining electrons.

The half-equation for the conversion of bromide ions into bromine molecules is:

$2Br^- \rightarrow Br_2 + 2e^-$

This is an oxidation half-equation as the bromide ions are losing electrons.

Exam tip

You may be asked to write individual half-equations for the conversion of molecules into ions or from ions into molecules. You may also have to identify these reactions as oxidation or reduction processes.

Knowledge check 7

Explain whether the half-equation $Cl_2 + 2e^- \rightarrow 2Cl^-$ is an oxidation or a reduction reaction, in terms of electrons.

Trends in reducing ability of the halide ions

Reducing agents are electron donors and they are oxidised in their reactions. The halide ions are reducing agents. The reducing ability of the halide ions increases down the group. Chloride ions are not as powerful a reducing agent as bromide ions, which in turn are not as powerful a reducing agent as iodide ions. This is because as you go down the group the ionic radius and shielding increase, and the attraction between the nucleus and the electron is reduced.

Reactions of solid sodium halides with concentrated sulfuric acid

Solid halides react with concentrated sulfuric acid. The concentrated sulfuric acid acts as an oxidising agent, and the halide ions are reducing agents. The equations (full and ionic), observations and names of the products of these reactions must be learnt. You must also be able to explain the redox reaction that is occurring.

Exam tip

Iodide ions are stronger reducing agents than chloride ions. They have a larger ionic radius and there is more shielding, so the electron lost from an iodide ion is less strongly held by the nucleus compared with the electron lost from a chloride ion.

Exam tip

There are eight equations, but the first equation in each is the same, just with a different halide ion and hydrogen halide. The second equation for bromide and iodide is also the same. Again, just change the halide and hydrogen halide. This means that you only have to remember four equations, rather than eight.

1 Reaction of concentrated H_2SO_4 with $NaF(s)$:

$NaF + H_2SO_4 \rightarrow NaHSO_4 + HF$ (not redox)
- Product names: sodium hydrogen sulfate and hydrogen fluoride.
- Observations: misty white fumes (HF).

The simplest ionic equation for this reaction is:

$F^- + H_2SO_4 \rightarrow HSO_4^- + HF$

Fluoride ions are not strong enough reducing agents to reduce the sulfur in sulfuric acid.

2 Reaction of concentrated H_2SO_4 with $NaCl(s)$:

$NaCl + H_2SO_4 \rightarrow NaHSO_4 + HCl$ (not redox)
- Product names: sodium hydrogen sulfate and hydrogen chloride.
- Observations: misty white fumes (HCl).

The simplest ionic equation for this reaction is:

$Cl^- + H_2SO_4 \rightarrow HSO_4^- + HCl$

Chloride ions are also not strong enough reducing agents to reduce the sulfur in sulfuric acid.

3 Reaction of concentrated H_2SO_4 with $NaBr(s)$:

$NaBr + H_2SO_4 \rightarrow NaHSO_4 + HBr$ (not redox)

$2HBr + H_2SO_4 \rightarrow Br_2 + SO_2 + 2H_2O$ (redox)
- Product names: sodium hydrogen sulfate, hydrogen bromide, bromine, water, sulfur dioxide.
- Observations: misty white fumes (HBr); red-brown vapour (Br_2).
- Redox: in the second equation the Br is oxidised from oxidation state -1 (in HBr) to 0 (in Br_2) and the S is reduced from +6 (in H_2SO_4) to +4 (in SO_2).

The bromide ion in hydrogen bromide is a better reducing agent than a chloride ion, and is a strong enough reducing agent to reduce the sulfur in sulfuric acid from the oxidation state of +6 in H_2SO_4 to +4 in SO_2.

The simplest ionic equations are:

$Br^- + H_2SO_4 \rightarrow HSO_4^- + HBr$

$2Br^- + SO_4^{2-} + 4H^+ \rightarrow Br_2 + SO_2 + 2H_2O$

4 Reaction of concentrated H_2SO_4 with $NaI(s)$:

$NaI + H_2SO_4 \rightarrow NaHSO_4 + HI$ (not redox)

$2HI + H_2SO_4 \rightarrow I_2 + SO_2 + 2H_2O$ (redox)

$6HI + H_2SO_4 \rightarrow 3I_2 + S + 4H_2O$ (redox)

$8HI + H_2SO_4 \rightarrow 4I_2 + H_2S + 4H_2O$ (redox)
- Product names: sodium hydrogen sulfate, hydrogen iodide, iodine, water, sulfur dioxide, sulfur, hydrogen sulfide.
- Observations: misty white fumes (HI); purple vapour (I_2); rotten egg smell (H_2S); white solid (NaI) changes to grey–black solid (I_2); yellow solid formed (S). This reaction should be carried out in fume cupboard because hydrogen sulfide is toxic.

> **Exam tip**
>
> Similar reactions occur with other halides, for example, potassium halide and concentrated sulfuric acid.

- Redox: in the second equation the I is oxidised from −1 (in HI) to 0 (in I_2) and the S is reduced from +6 (in H_2SO_4) to +4 (in SO_2).

  In the third equation the I is oxidised from −1 (in HI) to 0 (in I_2) and the S is reduced from +6 (in H_2SO_4) to 0 (in S).

  In the fourth equation the I is oxidised from −1 (in HI) to 0 (in I_2) and the S is reduced from +6 (in H_2SO_4) to −2 (in H_2S).

Iodide ions in hydrogen iodide are the best reducing agents of the three halides, and are capable of reducing the sulfur in sulfuric acid to SO_2, and even reducing it further to S and H_2S. Note that the solid reduction product is sulfur and the solid oxidation product is iodine.

The ionic equations for the reaction are:

$$I^- + H_2SO_4 \rightarrow HSO_4^- + HI$$

$$2I^- + SO_4^{2-} + 4H^+ \rightarrow I_2 + SO_2 + 2H_2O$$

$$6I^- + 8H^+ + SO_4^{2-} \rightarrow 3I_2 + S + 4H_2O$$

$$8I + 10H^+ + SO_4^{2-} \rightarrow 4I_2 + H_2S + 4H_2O$$

The reactions of the halides with concentrated sulfuric can be used as a test for a halide that is an alternative to the test with test with silver nitrate (detailed below):

- Chlorides give hydrogen chloride gas when reacted with concentrated sulfuric acid, which can be tested using a glass rod dipped in concentrated ammonia solution to give white smoke.
- Bromides give bromine vapour when reacted with concentrated sulfuric acid, which can be seen as brown fumes.
- Iodides give iodine vapour when reacted with concentrated sulfuric acid, which can be seen as purple fumes and/or a grey solid.

Knowledge check 8

For the reaction of concentrated sulfuric acid with potassium bromide, write one equation that is not a redox reaction.

Test for halide ions

The method for testing for a halide ion is as follows:

1. Make a solution of the compound using dilute nitric acid.
2. Add silver nitrate solution and record the colour of the precipitate.
3. Add dilute or concentrated ammonia solution.

	With silver nitrate solution	Add dilute ammonia solution	Add concentrated ammonia solution
Chloride (Cl^-)	White precipitate	White precipitate dissolves to give a colourless solution	White precipitate dissolves to give a colourless solution
Bromide (Br^-)	Cream precipitate	Cream precipitate remains	Cream precipitate dissolves to give a colourless solution
Iodide (I^-)	Yellow precipitate	Yellow precipitate remains	Yellow precipitate remains; it is insoluble in concentrated ammonia solution

Table 3

Exam tip

Why is dilute nitric acid added? It removes other ions that would react with the silver ions in the silver nitrate solution. For example, carbonate ions (CO_3^{2-}), sulfite ions (SO_3^{2-}) and hydroxide ions (OH^-) would react with the acid and form precipitates with the silver(I) ions, which would interfere with the test, so they must be removed.

The colour of the precipitate gives the identity of the ions, but dilute ammonia solution and/or concentrated ammonia solution can be used to confirm the identity of the ion present, as some of the precipitates will redissolve. The precipitates that dissolve form a soluble complex ion, for example, $[Ag(NH_3)_2]^+$.

Ionic equations for the tests:

$$Ag^+(aq) + Cl^-(aq) \rightarrow AgCl(s) \text{ white precipitate}$$

$$Ag^+(aq) + Br^-(aq) \rightarrow AgBr(s) \text{ cream precipitate}$$

$$Ag^+(aq) + I^-(aq) \rightarrow AgI(s) \text{ yellow precipitate}$$

> **Exam tip**
>
> A solution containing a mixture of sodium chloride and sodium iodide would form a cream precipitate with silver nitrate solution (resulting from a mixture of the white precipitate and the cream precipitate). Adding dilute ammonia solution leaves a yellow precipitate, as the white precipitate of silver chloride will dissolve.

> **Exam tip**
>
> Silver nitrate does not form a precipitate with fluoride ions in solution, as silver fluoride is soluble in water. Silver nitrate cannot be used to test for fluoride ions.

> **Knowledge check 9**
>
> What is observed when silver nitrate solution and then dilute ammonia solution are added sequentially to a solution containing iodide ions?

Required practical 4

You will need to be able to identify halide ions in solution and all of the other ions listed in Table 4. You may be given an unknown sample of an inorganic solid and asked to plan tests to identify the cation and the anion present in the solid. It is important to be able to link the reagent used in the test with the ion being tested for and the expected result. Questions may ask for the description of a test for a particular ion (you would need to describe the test and the results observed), or you may be given the test and the result and be expected to identify the ion. An alternative question style might provide the ion and the result, and ask you to describe how the test is carried out, including any reagents required.

Ion	Test	Result for positive test
Carbonate ions (CO_3^{2-})	Add a few drops of dilute nitric acid and bubble any gas produced through limewater	Effervescence Limewater turns from colourless to milky, indicating carbon dioxide release
Sulfate ions (SO_4^{2-})	Add a few drops of acidified barium chloride solution	White precipitate
Ammonium ion (NH_4^+)	Warm with sodium hydroxide solution and test the gas released with damp pH paper and a glass rod dipped in concentrated hydrochloric acid	Pungent gas evolved (ammonia) pH paper turns blue White smoke (ammonium chloride): $NH_3 + HCl \rightarrow NH_4Cl$
Group 2 cations	Add sodium hydroxide solution until in excess	White precipitate that does not dissolve in excess sodium hydroxide solution indicates magnesium or calcium ions present

Table 4

Chlorine and chlorate(I)

Reaction of chlorine with cold, dilute, aqueous sodium hydroxide

Chlorine reacts with cold, dilute, aqueous sodium hydroxide to produce chlorate(I) ions, ClO^-.

Ionic equation: $2OH^- + Cl_2 \rightarrow Cl^- + ClO^- + H_2O$

Balanced symbol equation: $2NaOH + Cl_2 \rightarrow NaCl + NaClO + H_2O$

Observations: green gas (chlorine) reacts to form a colourless solution.

The IUPAC name for NaClO is sodium chlorate(I) because it contains the chlorate(I) ion, ClO^-. The oxidation number of chlorine in chlorate(I) is +1.

This is a redox reaction, as chlorine is oxidised from the oxidation state of 0 in Cl_2 to +1 in NaClO, and chlorine is also reduced from 0 in Cl_2 to −1 in NaCl.

Uses of the solution formed

The solution containing sodium chlorate(I) is used as bleach. The chlorate(I) ions are responsible for the bleaching action of the solution. The sodium chlorate(I) kills bacteria and other microorganisms.

Knowledge check 10

Name the products formed when chlorine reacts with cold, dilute, aqueous sodium hydroxide.

Reaction of chlorine and water

Reaction of chlorine and water to form chloride ions and chlorate(I) ions

Chlorine is slightly soluble in water and produces a pale green solution. Some chlorine reacts with the water to form a mixture of hydrochloric acid (HCl) and chloric(I) acid (HClO). Chloric(I) acid contains the chlorate(I) ion. An equilibrium is established.

$Cl_2(g) + H_2O(l) \rightleftharpoons HCl + HClO$

Chlorine is added to water to kill microorganisms. The HClO reacts with bacteria in the water and the position of equilibrium moves to the right to replace the HClO that has reacted. Hence, when the HClO has done its job, there is little chlorine left in the water.

Chlorine is used to sterilise drinking water and water in swimming pools. There are benefits of using chlorine in water supplies:
- It kills disease-causing microorganisms.
- It prevents the growth of algae and prevents bad taste and smells.
- It removes discolouration caused by organic compounds.

There are risks associated with using chlorine to treat water, as it is a toxic gas and it may react with any organic compounds that are present in water from the decomposition of plants, to form chlorinated hydrocarbons, which may cause cancer.

However, chlorine is used in water treatment in dilute concentrations, and when it has done its job, owing to the equilibrium reaction, there is little of it remaining in the water. The benefits to health of water treatment outweigh chlorine's toxic effects.

Exam tip

Questions may ask for this reaction specifically as an equilibrium reaction, so the reversible arrow will be expected in the answer. You must also be able to explain the redox — chlorine is both oxidised from 0 in Cl_2 to +1 in HOCl, and reduced from 0 in Cl_2 to −1 in HCl.

The reaction of chlorine and water to form chloride ions and oxygen

In sunlight the chloric(I) acid decomposes into hydrochloric acid. The equation is:

$$Cl_2 + 2H_2O \rightarrow 4HCl + O_2$$

This means that chlorine is rapidly lost from pool water in sunlight.

Summary

- The boiling points of the halogens increase down the group because of increasing van der Waals forces, and the electronegativity decreases down the group as the atomic radius and shielding increase.
- The halogens decrease in oxidising ability as the group is descended. This is shown in the ability of a more reactive halogen to displace a less reactive halogen from a solution of its halide ion.
- The reducing ability of halide ions increases down the group as a result of the increasing ionic radius and increasing shielding. The trend in reducing ability of halides is seen in the different reactions of solid halide with concentrated sulfuric acid.

- Nitric acid and silver nitrate solution can be used to test for halide ions. The colour of the precipitate and whether it redissolves in ammonia solution indicate which halide ion is present.
- Chlorine reacts with cold, dilute aqueous sodium hydroxide to produce sodium chloride, water and sodium chlorate(I), which is used in bleach.
- Chlorine reacts with water to produce HClO and HCl. However, in the presence of sunlight, HCl and O_2 are produced. Despite its toxicity, the benefits of adding chlorine to water to kill microorganisms outweigh the risks.

■ Introduction to organic chemistry

Organic chemistry is the study of the millions of covalent compounds of the element carbon. The study of organic chemistry is made simpler by arranging the compounds into families, or homologous series, which contain the same functional group. A **functional group** is a group of atoms that are responsible for the characteristic reactions of a molecule.

Formulae

Organic compounds can be represented by different types of formula.

Molecular formula

This shows the actual number of atoms of each element in a compound.

Compound	Molecular formula	Number of atoms present
Ethane	C_2H_6	Two atoms of C and six atoms of H
Chloropropane	C_3H_7Cl	Three atoms of C, seven atoms of H, one atom of Cl
Ethanoic acid	$C_2H_4O_2$	Two atoms of C, four atoms of H, two atoms of O

Table 5

Exam tip

The functional group is not shown in a molecular formula. Only the number of atoms of each element is given. For example, the molecular formula of ethanol is C_2H_6O, not C_2H_5OH, and the molecular formula of ethanoic acid is $C_2H_4O_2$, not CH_3COOH.

Empirical formula

This is the simplest whole number ratio of each element in a compound.

Compound	Molecular formula	Empirical formula
Ethane	C_2H_6	CH_3
Chloropropane	C_3H_7Br	C_3H_7Br
Ethanoic acid	$C_2H_4O_2$	CH_2O

Table 6

Notice that the empirical and molecular formulae of chloropropane are the same.

General formula

This is an algebraic formula, which can describe any member of a homologous series.

Homologous series	General formula (n = number of carbon atoms)
Alkanes	C_nH_{2n+2}
Alkenes	C_nH_{2n}
Halogenoalkanes	$C_nH_{2n+1}X$ (where X = a halogen)
Alcohols	$C_nH_{2n+1}OH$

Table 7

Worked example

What is the molecular formula of an alcohol with six carbons?

Answer

General formula is $C_nH_{2n+1}OH$; number of C = n = 6

Number of H = $2n + 1 = (2 \times 6) + 1 = 12 + 1 = 13$

The formula is $C_6H_{13}OH$.

Knowledge check 11

What is the molecular formula of an alkane containing 10 carbon atoms?

Displayed formula

This shows how all the atoms and all the bonds between them are arranged. Ionic parts of the molecule are shown using charges, as shown in Figure 3.

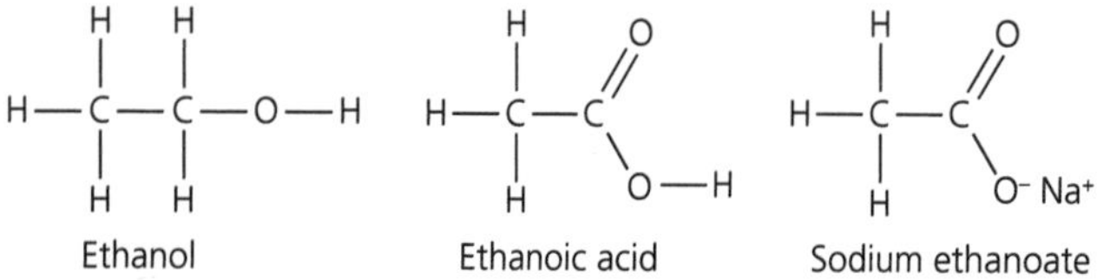

Ethanol   Ethanoic acid   Sodium ethanoate

Figure 3

Structural formula

This shows the arrangement of atoms in a molecule, carbon by carbon, with the attached hydrogens and functional groups, but without showing the bonds. Brackets are used to indicate that a group is bonded to the previous carbon atom and is not part of the main chain. For example, butanoic acid has the displayed formula shown in Figure 4, and the structural formula $CH_3CH_2CH_2COOH$.

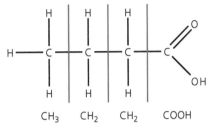

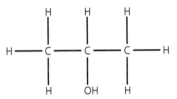

Figure 4

Propan-2-ol has the displayed formula shown in Figure 5. Its structural formula is $CH_3CH(OH)CH_3$.

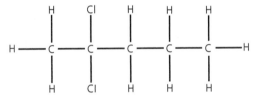

Figure 5

The displayed formula of 2,2-dichloropentane is shown in Figure 6. Its structural formula is $CH_3CCl_2CH_2CH_2CH_3$.

Figure 6

The displayed formula of 3,3-dimethylpentane is shown in Figure 7. Its structural formula is $CH_3CH_2C(CH_3)_2CH_2CH_3$.

Figure 7

Exam tip

Identical adjacent groups can be combined. For example, the two CH_2 groups in the middle of the butanoic acid molecule (Figure 4) can be combined to give the condensed structural formula, $CH_3(CH_2)_2COOH$.

Exam tip

The OH group is shown in brackets because it is not part of the main chain and is bonded to a carbon atom in the middle of the chain.

Exam tip

Single atoms do not require brackets if more than one is bonded to a carbon atom.

Exam tip

Sometimes the C=C in alkenes is shown in structural formulae — for example, $CH_2{=}CH_2$ instead of CH_2CH_2 for ethene.

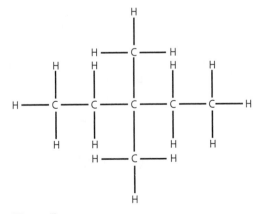

Write the molecular, empirical and structural formulae for the molecule shown in Figure 8.

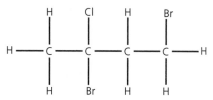

Figure 8

Skeletal formula

This shows just the carbon skeleton, with hydrogen atoms removed and functional groups present. Each carbon–carbon bond is shown as a line.

In a skeletal formula:

■ There is a carbon atom at each junction between bonds in a chain and at the end of each bond (unless there is something else there already — like the –OH group in an alcohol).

■ There are hydrogen atoms attached to each carbon to make the total number of bonds on that carbon up to four.

Name	Skeletal formula	Structural formula
Butan-1-ol		$CH_3CH_2CH_2CH_2OH$
Butan-2-ol		$CH_3CH(OH)CH_2CH_3$
Pent-1-ene		$CH_2=CH_2CH_2CH_2CH_3$
Methanoic acid		HCOOH
Ethanoic acid		CH_3COOH
Cyclopentane		C_5H_{10}

Table 8

Note that for the skeletal formula of methanoic acid a hydrogen atom must be placed at the end of the chain to show that there is only one carbon in the molecule.

Write the skeletal formula of 1-chloropropane.

Nomenclature

The rules for nomenclature (naming) of organic compounds are based on the IUPAC (International Union of Pure and Applied Chemists) system. The correct chemical name of a compound is often called the IUPAC name. You must be able to use IUPAC rules to name organic compounds — either rings or chains with up to six carbon atoms. A name is made up of a prefix, a stem and a suffix. You need to understand the following rules, and be able to apply them in the examples.

1   Count the number of carbon atoms in the longest unbranched chain and find the **stem** part of the name using Table 9.

Number of carbon atoms	Stem
1	Meth
2	Eth
3	Prop
4	But
5	Pent
6	Hex

Table 9

Exam tip

In exam questions, watch out for the longest chain not being drawn completely straight — this is done to try to catch you out! Always count the longest continuous chain.

2   Identify any side groups and name them using a **prefix**. A prefix is added before the stem as part of the name. Some common prefixes are shown in Table 10.

Side group or substituent	Prefix
$-CH_3$	Methyl
$-CH_2CH_3$	Ethyl
$-CH_2CH_2CH_3$	Propyl
$-F$	Fluoro
$-Cl$	Chloro
$-Br$	Bromo
$-I$	Iodo

Table 10

Exam tip

If the compound has the carbon atoms arranged in a ring, then the prefix 'cyclo' is added to the name. Cyclohexane is shown in Figure 9.

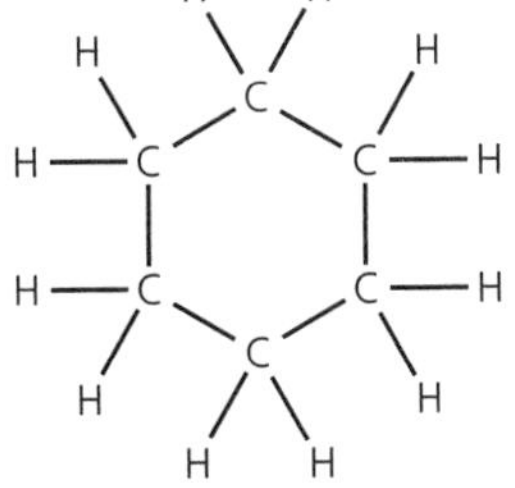

Figure 9

3 The suffix depends on the functional group. Alkanes are the simplest compounds to name — the **suffix** is '-ane'. Alkenes have a double bond and the suffix '-ene'. Alcohols have the suffix '-ol'.

4 All substituent groups are named alphabetically, for example, chloro comes before methyl.

5 If there is more than one of the same substituent group, this is prefixed with 'di-', 'tri-', 'tetra-' and so on. For example, dichloro if there are two chlorine atoms (even if they are bonded to different carbon atoms), and trimethyl if there are three methyl groups (even if each is bonded to a different carbon atom).

6 Each substituent group must have a number to indicate its position on the carbon chain. This is often called a locant number and is placed in front of the substituent group. A separate number is needed for each substituent. Commas are used to separate the numbers. Dichloro requires two locant numbers, one for the position of each chlorine atom, for example 1,2-dichloro. The carbon atoms in the longest chain are numbered from the end that gives the lowest locant numbers.

7 Dashes are placed between numbers and letters.

> ### Exam tip
>
> The di-, tri- and tetra-prefixes do not change the alphabetical order — this is based on the name of the substituent group. For example, trichloro, difluoro is a correct order, based on the name of the substituent group, not the prefix.

Worked example 1

Give the IUPAC name for the structure in Figure 10.

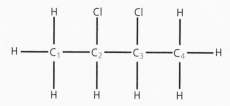

Figure 10

Answer

Name the longest unbranched carbon chain — there are four carbons, so the stem name is 'but'.

There are two chloro side groups, so the prefix is 'dichloro'.

The chloro groups are on carbon 2 and carbon 3.

The name is 2,3-dichlorobutane.

> ### Exam tip
>
> Number the carbon atoms — as shown in Figure 10 in red. It makes it easier to see where the side groups are attached. It also helps to circle the side groups.

Worked example 2

Give the IUPAC name for the structure in Figure 11.

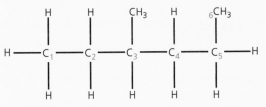

Figure 11

Answer

Name the longest unbranched carbon chain —there are six carbons, so the stem name is 'hex'.

There is one CH_3 side group so the prefix is 'methyl'.

The methyl group is on carbon 3.

The name is 3-methylhexane.

Worked example 3

Give the IUPAC name for the structure in Figure 12.

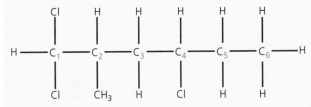

Figure 12

Answer

Name the longest unbranched carbon chain — there are six carbons, so the stem name is 'hex'.

There is one CH_3 side group, so the prefix is 'methyl', and there are three chlorines attached, which is 'trichloro'.

The methyl group is on carbon 2 and the chloros are on carbons 1 and 4.

The name is 1,1,4-trichloro-2-methylhexane. Note the alphabetical order of substituents.

Exam tip

In worked example 2, watch out for the methyl group that is bonded to the right-hand carbon in the straight part of the chain — it is part of the longest unbranched carbon chain.

Knowledge check 14

Give the IUPAC name of $CH_3CBr_2CH_2CH_3$.

Naming molecules with functional groups

Common examples of molecules that contain a functional group are given in Table 11, in order of decreasing nomenclature priority.

	Homologous series	Functional group	Nomenclature style and example
Highest priority	Carboxylic acid	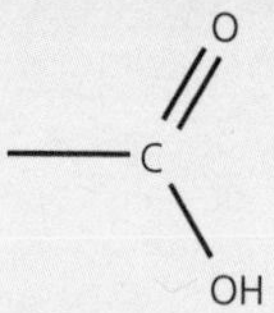	-oic acid e.g. ethanoic acid
	Ester	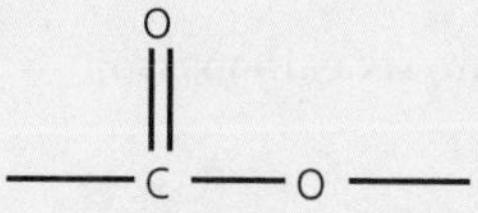	Alkyl carboxylate e.g. ethyl ethanoate
	Acid chloride	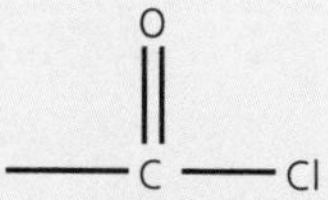	Alkanoyl chloride e.g. ethanoyl chloride
	Nitrile	—C≡N	-nitrile e.g. ethanenitrile
	Aldehyde	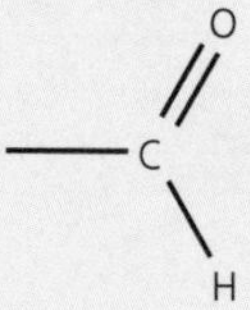	-al e.g. propanal
	Ketone	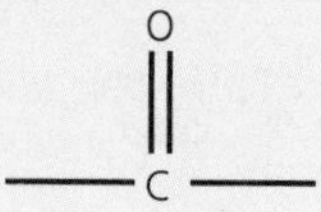	-one e.g. butanone
	Alcohol	—OH	-ol propan-1-ol
	Amine	—NH_2	-amine e.g. ethylamine
	Alkene	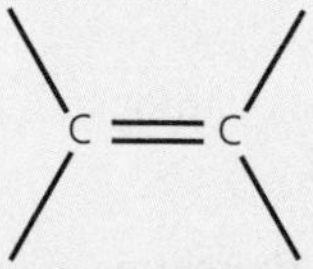	-ene e.g. propene
Lowest priority	Halogenoalkane	—C—X	Named as a substituted hydrocarbon e.g. chloroethane

Table 11

An alcohol has the suffix '-ol', but if a higher priority group is also present — e.g. if there is an aldehyde and an OH in one molecule — the molecule is named '-al' for the aldehyde and the OH is named by the prefix 'hydroxy'.

The priority rule is important when a molecule has two or more functional groups. The highest priority group becomes the main name and the lower priority group(s) are named as substituent groups. Anything with lower priority than an alkene is named as a substituent group, for example, chloro, bromo, iodo (and methyl).

Exam tip

Even though the term halogenoalkane contains the word 'alkane', the halogen atom is a functional group because it determines the characteristic reactions of the compound.

If the suffix starts with a vowel, then remove the final letter 'e' from the alkane part of the name. For example, 'methaneal' becomes 'methanal'.

If the suffix starts with a consonant, then the 'e' is kept. For example, ethanedioic acid.

Worked example 1

Give the IUPAC name of the structure in Figure 13.

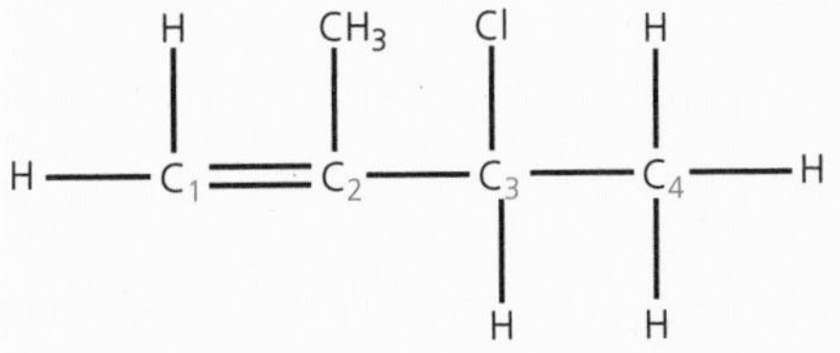

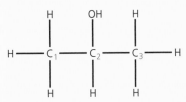

Figure 13

Answer

Name the longest unbranched carbon chain — there are four carbons, so the stem name is 'but'.

There is one CH_3 side group, so the prefix is 'methyl', and one chloro is also attached, which is 'chloro'.

The methyl group is on carbon 2 and the chloro is on carbon 3. The position of the double bond must be stated. This applies for alkenes with four or more carbons in the chain. The position at which the double bond starts is given — in this case carbon 1.

The name is 3-chloro-2-methylbut-1-ene.

Worked example 2

Give the IUPAC name for the structure in Figure 14.

H —— C$_1$ —— C$_2$ —— C$_3$ —— H

Figure 14

Answer

Name the longest unbranched carbon chain — there are three carbons, so the stem name is 'prop'.

There is one –OH functional group, so the suffix is '-ol'.

The position of the –OH must be stated — it is on carbon 2. This applies for alcohols with three or more carbons in the chain.

The name is propan-2-ol.

Exam tip

The suffix starts with a vowel so remove the final 'e' from the alkane chain — propane becomes 'propan'.

Worked example 3

Give the IUPAC name for the structure in Figure 15.

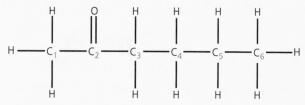

Figure 15

Answer

Name the longest unbranched carbon chain — there are three carbons, so the stem name is 'prop'.

There is one –OH functional group so the suffix is '-ol'.

There is one CH_3 side group, which is 'methyl'.

The position of the –OH and the methyl must be stated. Remember to use the lowest number locants. (Numbering the carbons in the opposite direction would give the positions 2 and 3, which is a higher sequence of numbers and is incorrect.)

The name is 2-methylpropan-1-ol.

Worked example 4

Give the IUPAC name for the structure in Figure 16.

```
    H     O     H     H     H     H
    |     ‖     |     |     |     |
H — C₁ — C₂ — C₃ — C₄ — C₅ — C₆ — H
    |           |     |     |     |
    H           H     H     H     H
```

Figure 16

Answer

Name the longest unbranched carbon chain — there are six carbons, so the stem name is 'hex'.

There is one C=O functional group in the chain, the compound is a ketone, so the suffix is '-one'.

The position of the C=O must be stated — it is on carbon 2. This applies for ketones with five or more carbons in the chain.

The name is hexan-2-one.

Knowledge check 15

Give the IUPAC name for:

a $CH_3CHClCH_2OH$

b $CH_3CH(CH_3)CHO$

Worked example 5

Give the IUPAC name for the structure in Figure 17.

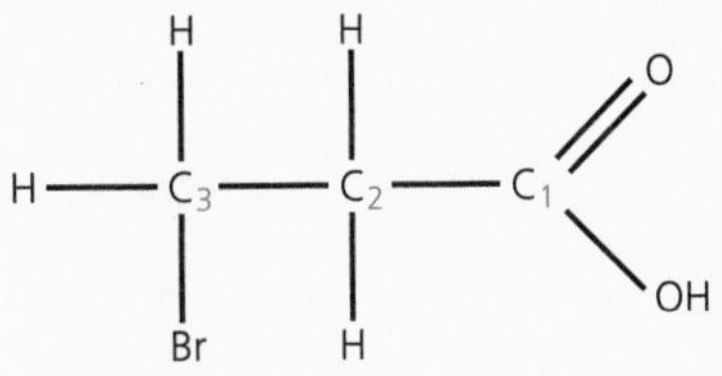

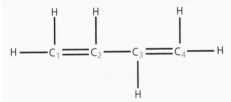

Figure 17

Answer

Name the longest unbranched carbon chain — there are three carbons, so the stem name is 'prop'.

There are two functional groups — one –COOH and one Br group. The highest priority group is the COOH, so it is named as a carboxylic acid.

The chain is numbered from the highest priority group, so the bromo group is on carbon 3.

The name is 3-bromopropanoic acid.

Exam tip

Misplacing commas and dashes will lose marks. Make sure that numbers are separated by commas and that you put a dash between numbers and letters.

Sometimes a molecule has two identical functional groups.

Worked example 6

Give the IUPAC name for the structure in Figure 18.

Figure 18

Answer

Name the longest unbranched carbon chain — there are four carbons, so the stem name is 'but'.

There are two double bonds, so the suffix becomes '-diene'.

The position of the double bonds must be given, which is 1,3.

The stem of the name gets an extra 'a', which gives 'buta'.

The name is buta-1,3-diene.

Isomers

Isomers occur when molecules with the same molecular formula have a different arrangement of atoms. There are two main types of isomerism:

- structural isomerism
- stereoisomerism

Structural isomers

Structural isomers are compounds that have the same molecular formula but different structural formulae. There are three types of structural isomer.

Chain isomers

These occur when there is more than one way of arranging the carbon chain of a molecule — for example, the carbon skeleton of an alkane of formula C_4H_{10} can be drawn as a straight chain or as a branched chain (Figure 19).

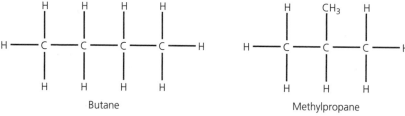

Butane   Methylpropane

Figure 19

C_5H_{12} has three chain isomers, as shown in Figure 20.

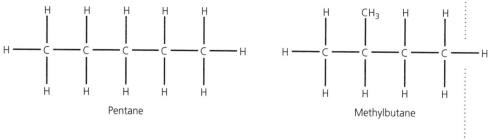

Pentane   Methylbutane

Dimethylpropane

Figure 20

> **Exam tip**
>
> Methylpropane is often called 2-methylpropane, even though there is no other position for the methyl group. Methylbutane is often called 2-methylbutane, and dimethylpropane can be called 2,2-dimethylpropane. In an exam, there would be no penalty for including the locant numbers.

C_6H_{14} has five chain isomers, as shown in Figure 21.

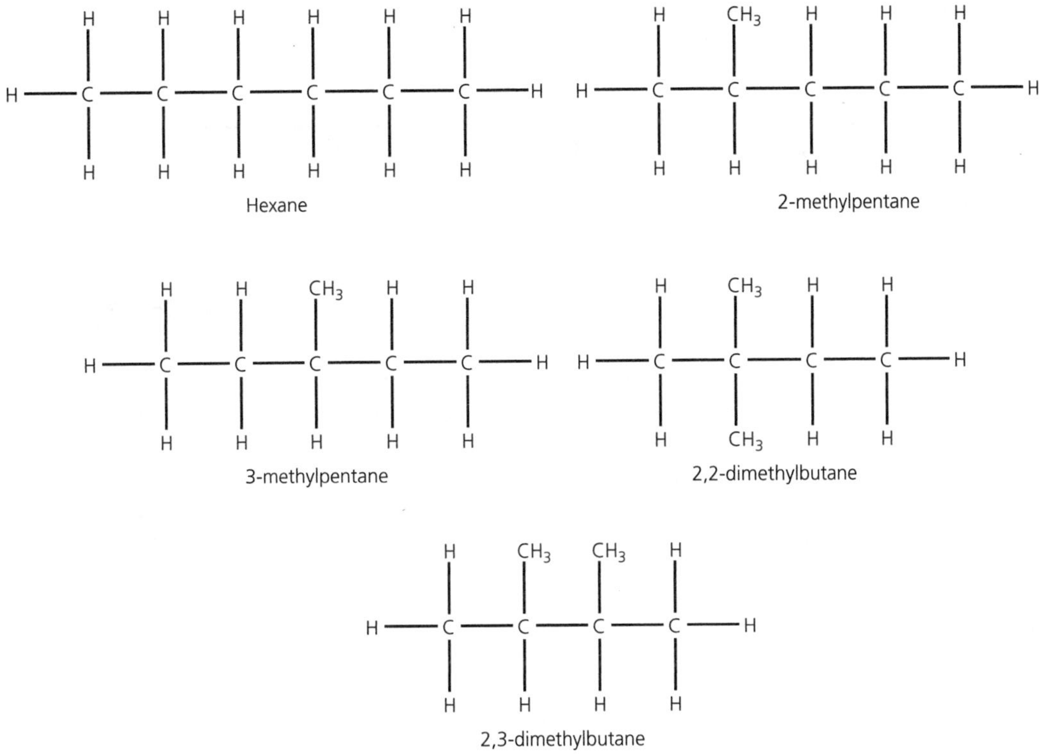

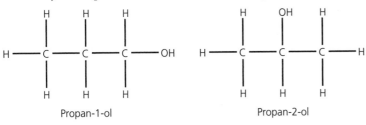

Hexane

2-methylpentane

3-methylpentane

2,2-dimethylbutane

2,3-dimethylbutane

Figure 21

Exam tip

If you are asked to draw several isomers of a formula, always name the isomers. This helps you to avoid drawing the same structure twice. It is also helpful to start with the longest carbon chain, and then to shorten the chain and move the position of the alkyl group to obtain other isomers.

Chain isomers have similar chemical properties but slightly different physical properties. The more branched the isomer, the weaker the van der Waals forces between the molecules and the lower the boiling point.

Position isomers

Position isomers have the same carbon chain, but the functional group is bonded to different carbons. The two position isomers of C_3H_7OH are shown in Figure 22. They differ only in the position of the –OH functional group.

Propan-1-ol

Propan-2-ol

Figure 22

Knowledge check 16

Name two chain isomers of $C_4H_8O_2$.

Content Guidance

The molecules shown in Figure 23 (pentan-2-one and pentan-3-one) differ only in the position of the ketone carbonyl group. They are position isomers.

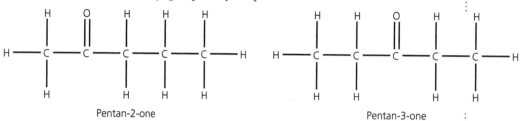

Figure 23

Functional group isomers

Functional group isomers have the same molecular formula but different functional groups. The molecules shown in Figure 24 have different functional groups, yet they have the same molecular formula ($C_4H_8O_2$). Ethyl ethanoate is an ester and butanoic acid is a carboxylic acid. They are functional group isomers.

Ethyl ethanoate

Butanoic acid

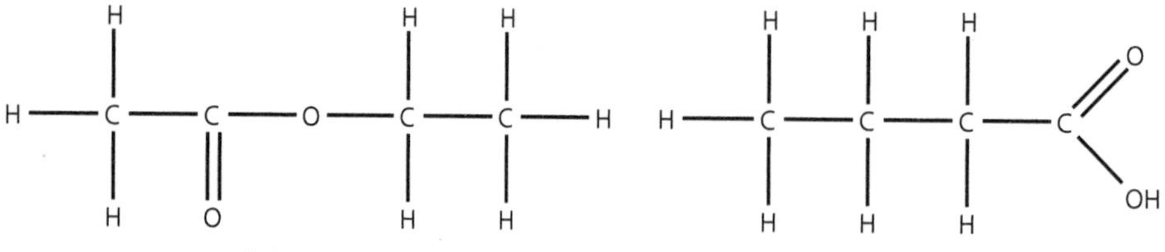

Figure 24

Cycloalkanes are functional group isomers of alkenes. For example, hexene has a C=C functional group and the molecular formula C_6H_{12}. Cyclohexane also has the molecular formula C_6H_{12}, but it is an alkane and does not have the C=C functional group (Figure 25).

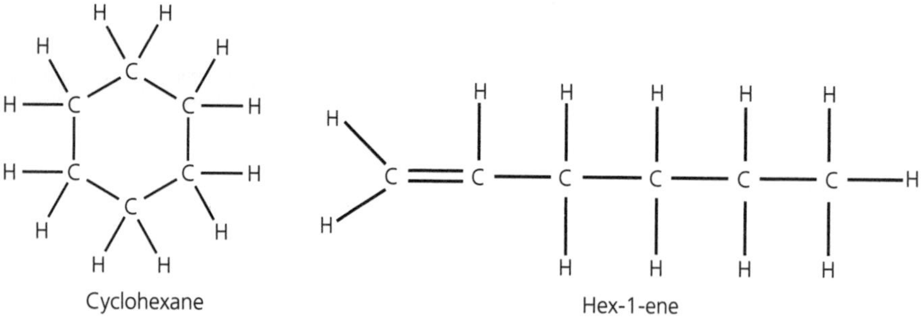

Cyclohexane

Hex-1-ene

Figure 25

The molecules shown in Figure 26 also have different structures, but the same molecular formula (C_3H_6O). Propanone is a ketone and propanal is an aldehyde.

Knowledge check 17

Give the IUPAC name of:

a the position isomer
b the chain isomer

of but-1-ene.

Knowledge check 18

Give the IUPAC name of a position isomer of hex-1-ene.

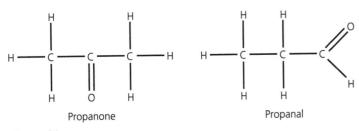

Propanone   Propanal

Figure 26

Stereoisomers

Stereoisomers have the same structural and molecular formula, but a different arrangement of atoms in three-dimensional (3D) space. There are two types of stereoisomers: E–Z isomers and optical isomers (studied in year 2).

E–Z isomers

A single carbon–carbon bond allows free rotation. However, there is an energy barrier to free rotation about a planar carbon–carbon double bond, and it is this restricted rotation that leads to E–Z isomerism in some alkenes.

The criteria for E–Z isomers are:

- a carbon–carbon double bond must be present, and
- each carbon in the double bond must be attached to two different groups

Hence, but-1-ene does not have E–Z isomers, as the two groups on the first carbon (circled in blue in Figure 27) are the same — they are both H. In contrast, but-2-ene has E–Z isomers, as each carbon in the double bond is attached to two different groups — an H and a CH_3.

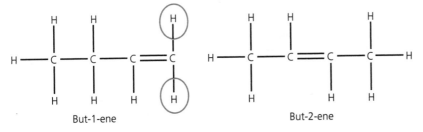

But-1-ene   But-2-ene

Figure 27

To identify E–Z isomers:

- Look at the first carbon in the double bond and decide on the priority of the groups attached to it. This is done using Cahn–Ingold–Prelog (CIP) priority rules — the higher the atomic number of the atom attached directly to the carbon of the C=C, then the higher the priority. If two atoms are the same, consider the total atomic number of the atoms bonded directly to them. Remember, atoms attached by double bonds have their atomic number counted twice.
- If both of the substituents of higher priority are on the same side of the plane of the C=C bond, the arrangement is Z. If they are on opposite sides, the arrangement is E.

Exam tip

There are three classes of organic compound that are functional group isomers of each other that you will come across during your course:

- Esters are functional group isomers of carboxylic acids.
- Cycloalkanes are functional group isomers of alkenes.
- Aldehydes are functional group isomers of ketones.

Exam tip

Z is from the German word *zusammen*, which means *together*. E is from the German word *entgegen*, which means *opposite*.

Content Guidance

The E–Z isomers of but-2-ene are shown in Figure 28. On each carbon the two atoms attached to the carbon of the double bond are H (atomic number 1) and C (atomic number 6). Carbon has the highest atomic number and highest priority and so in the Z isomers, both methyl groups are on the same side of the carbon–carbon double bond, but in the E isomer they are on opposite sides.

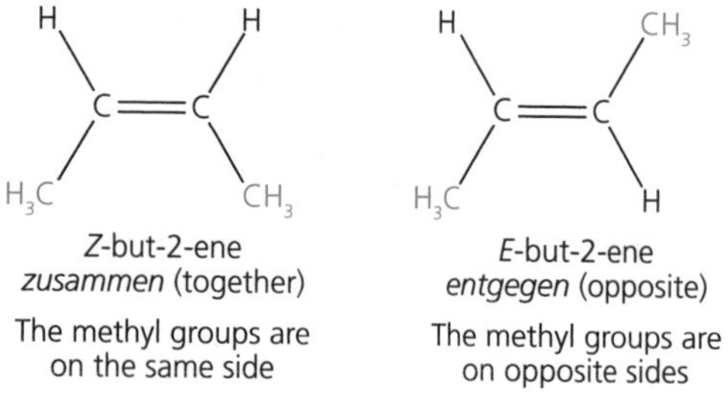

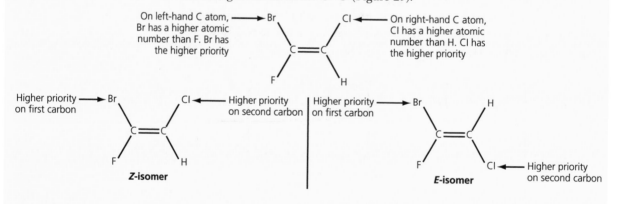

Z-but-2-ene
zusammen (together)

The methyl groups are
on the same side

E-but-2-ene
entgegen (opposite)

The methyl groups are
on opposite sides

Figure 28

Worked example 1

Draw the E–Z isomers of 1-bromo-2-chloro-1-fluoroethene.

Answer

First, draw the structural formula and then consider the priority of the groups bonded to the left carbon and then the right carbon of the C=C (Figure 29).

On left-hand C atom, —→ Br
Br has a higher atomic
number than F. Br has
the higher priority

Cl ←— On right-hand C atom,
Cl has a higher atomic
number than H. Cl has
the higher priority

Higher priority —→ Br
on first carbon

Cl ←— Higher priority
on second carbon

Higher priority —→ Br
on first carbon

H

Z-isomer

E-isomer

Cl ←— Higher priority
on second carbon

Figure 29

Exam tip

The isomers are named by placing a capital E or Z followed by a dash and then the name, for example, E-1-bromo-2-chloro-1-fluoroethene. Note that sometimes a question asks you to give the full IUPAC name for a stereoisomer, so you need to include the E or Z in the name.

Worked example 2

Classify the isomers shown in Figure 30 as *E* or *Z*.

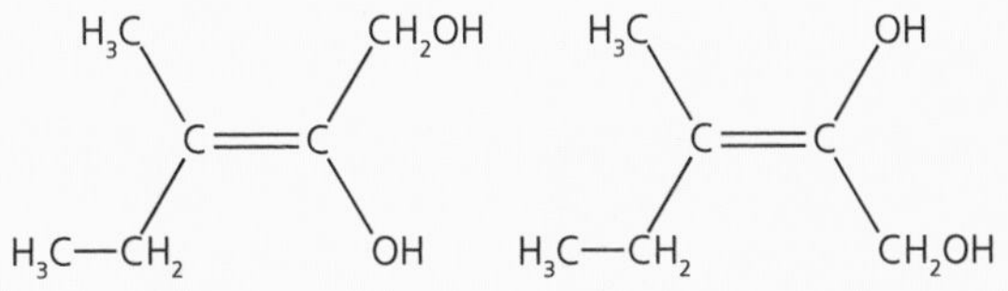

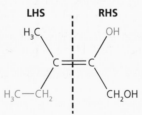

Figure 30

Answer

When the molecule contains groups rather than single atoms, the system works in a similar way. First, look at the atoms that are bonded directly to each carbon in the double bond. Then, move one atom away from the double bond and look at the atoms bonded there (Figure 31).

The top group has a C atom bonded directly to the C=C, atomic number 6. The bottom group also has a carbon directly bonded to the C=C. This is the same priority. Now look at all the atoms bonded directly to these carbons. The top carbon has 3 H atoms. This is a total atomic number of 3. The bottom carbon has 2 H atoms and a C atom. The total atomic number is 8. Therefore the bottom group has the higher priority.

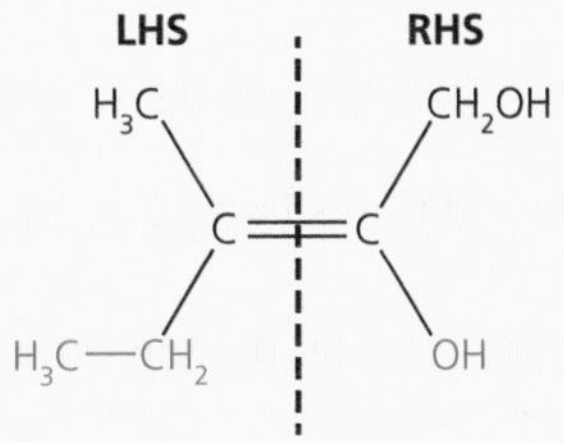

LHS | **RHS**

The top group has a C atom bonded directly to the C=C, atomic number 6. The bottom group has an oxygen atom bonded directly to the C=C, atomic number 8. The OH group has the higher priority.

The higher priority groups are on the same side of the plane of the double bond.

Z-2-hydroxy-3-methylpent-2-en-1-ol

As before, the atoms directly attached to the C atom of the C=C are the same and the priority is decided by the atoms next along the chain.

LHS | **RHS**

As before the OH group has the higher priority as it has the atom with the higher atomic numer bonded to the C atom in the C=C.

The higher priority groups are on the opposite side of the plane of the double bond.

E-2-hydroxy-3-methylpent-2-en-1-ol

Figure 31

Worked example 3

Classify the isomer shown in Figure 32 as *E* or Z. Explain your reasoning using CIP rules.

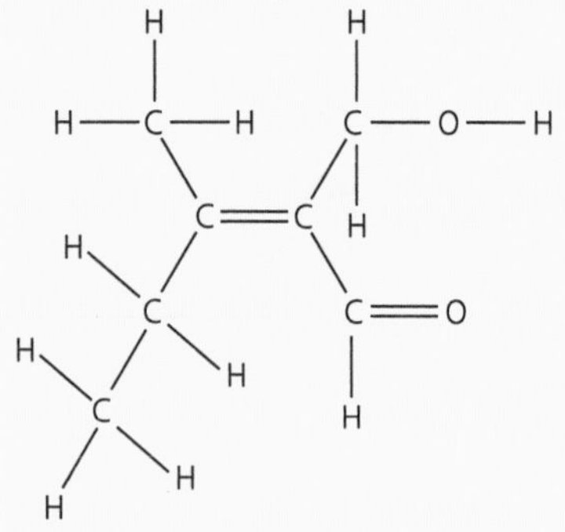

Figure 32

Answer

Left-hand carbon:

- On the left-hand carbon of the double bond there are two carbons attached, of same atomic number and priority.
- Moving to the atoms attached to these carbons: on the top carbon there are 3H attached — atomic number total of 3; and on the bottom carbon there is a C and 2H attached — atomic number total of 8, so the bottom group has higher priority.

Right-hand carbon:

- On the right-hand carbon of the double bond there are two carbons, of the same atomic number and priority.
- Moving to the atoms attached to these carbons: on the top carbon there are two H and one O attached, which gives an atomic number total of 10. On the bottom carbon there is 1H and 1O directly attached, but the O is attached by a double bond, and so its atomic number must be added twice, so the atomic number total is $(1 + (2 \times 8)) = 17$. Hence, the CHO group has a higher priority.

The highest priority groups are on the same side of the double bond, so the isomer shown is the Z isomer.

Knowledge check 19

What is the origin of *E–Z* isomerism?

Reaction mechanisms

Reactions of organic compounds can be explained using mechanisms. You will study mechanisms for the reactions of many different homologous series — these mechanisms will be described in the sections relevant to each homologous series.

Summary

- There are six main types of formulae used in organic chemistry — molecular, empirical, general, displayed, structural and skeletal.
- Nomenclature of alkanes, substituted alkanes and halogenoalkanes is based on the longest continuous alkane chain. Locant numbers are used to identify the positions of any substituent groups (if necessary).
- Nomenclature of molecules from other homologous series is based on a terminal name ending that depends on the functional group(s) present.

- Structural isomers are compounds that have the same molecular formula but a different structural formula. There are three types of structural isomers — chain, position and functional.
- Stereoisomers have the same structural and molecular formulae but a different arrangement of atoms in three-dimensional space. E–Z isomers are stereoisomers. In the E isomer the highest priority groups are on opposite sides of the double bond, whereas in the Z isomer they are on the same side.

■Alkanes

Fractional distillation of crude oil

Alkanes are a **homologous series** with the general formula C_nH_{2n+2}. Alkanes are **saturated hydrocarbons**. Petroleum is the raw material from which alkanes are obtained.

- Petroleum (crude oil) is a mixture that consists mainly of alkane hydrocarbons. The separation of hydrocarbons in crude oil is carried out by **fractional distillation**. Alkanes are a major product.
- Petroleum is separated into a number of simpler mixtures (fractions) containing molecules that have different boiling point ranges.
- Petroleum vapour is passed into a fractionating column, which has a temperature gradient — it is hot at the bottom and cooler at the top. As the vapour moves up the column it gets cooler, and owing to the different chain lengths and boiling points of different hydrocarbons, each fraction condenses at a different temperature. The fractions are drawn off at different levels in the column.
- The larger hydrocarbons present in the petroleum mixture do not vaporise because their boiling points are too high, and they are collected at the bottom of the column.
- The smaller hydrocarbons with the lowest boiling points do not condense, and they exit the column at the top as gases.
- Fractions such as petrol, diesel, kerosene and lubricating oil are produced from the fractional distillation of petroleum.

Exam tip

The smaller hydrocarbon molecules have smaller M_r and fewer electrons, and so weak van der Waals forces of attraction between their molecules, which do not take much energy to break. Hence, they have low boiling points and condense higher in the column where it is cooler.

A **homologous series** is a family of organic compounds that possess the same general formula, show similar chemical properties and show a gradual change in physical properties as the homologous series is ascended.

Hydrocarbons are compounds that contain carbon and hydrogen atoms only.

Saturated means that the compound only contains single C–C bonds — there are no C=C bonds.

Fractional distillation is the separation of components in a mixture based on their different boiling points. The technique involves boiling and condensing, and collecting fractions over particular boiling point ranges.

Modification of alkanes by cracking

In the fractional distillation of petroleum the longer-chain fractions, for example, lubricating oil, are not as useful as the shorter-chain fractions (light fractions). Demand for shorter-chain fractions, which are more useful as fuels and for organic synthesis, exceeds supply. To meet this demand, **cracking** is used to break heavier long chain molecules into more useful, small chain hydrocarbons.

A general equation to describe cracking is:

$$C_xH_y \rightarrow C_aH_b + C_cH_d$$

where $x = a + c$ and $y = b + d$.

The numbers of carbon and hydrogen atoms on the left- and right-hand sides of the equation have to balance. At least one product must be an alkene (unsaturated). For example:

$$C_{18}H_{38} \rightarrow C_{10}H_{22} + C_8H_{16}$$

saturated   saturated   unsaturated

Another example of thermal cracking has the equation:

$$C_{18}H_{38} \rightarrow C_8H_{18} + C_4H_8 + C_6H_{12}$$

saturated   saturated   unsaturated   unsaturated

You may be asked to write equations for cracking. Usually the question gives a lot of information about the products.

Worked example

A hydrocarbon is cracked to produce 2 moles of ethene, 1 mole of but-1-ene and 1 mole of octane. Write a balanced symbol equation for the reaction.

Answer

Allow C_xH_y to be the hydrocarbon and then write down the formula of the products using the information in the question:

$$C_xH_y \rightarrow C_2H_4 + C_4H_8 + C_8H_{18}$$

Then insert the number of moles given in the question:

$$C_xH_y \rightarrow 2C_2H_4 + C_4H_8 + C_8H_{18}$$

Remember that the equation must balance, so the number of the carbons and hydrogens on both sides must be equal.

$$x = (2 \times 2) + 4 + 8 = 16$$

$$y = (2 \times 4) + 8 + 18 = 34$$

$$C_{16}H_{34} \rightarrow 2C_2H_4 + C_4H_8 + C_8H_{18}$$

Knowledge check 20

Camping gas contains liquefied C_3H_8 and C_4H_{10}. Explain which compound liquefies most easily.

Cracking is the breakdown of large alkanes into smaller molecules by breaking the C–C bonds. It requires high temperatures to break the strong C–C bond in alkanes.

Exam tip

The main purpose of cracking is to produce products that are in greater demand, as they are more useful and have a higher value.

Exam tip

You can tell whether a compound is unsaturated or not by the formula. If a formula follows C_nH_{2n+2}, then the hydrocarbon is an alkane and is saturated. If the formula follows C_nH_{2n}, then the hydrocarbon is an alkene and is unsaturated.

Knowledge check 21

Write an equation for the thermal cracking of octane, C_8H_{18}, to form ethene and propene, and one other product. Name the other product.

There are two different methods of cracking used in industry, which employ different conditions and generate different types of products, as shown in Table 12.

	Thermal cracking	Catalytic cracking
Temperature	High (1000°C)	High (450°C)
Pressure	High	Slight
Catalyst	None	Zeolite
Products	High percentage of alkenes (which are useful to make polymers)	Motor fuels and aromatic hydrocarbons

Table 12

Combustion of alkanes

Alkanes are used as fuels, as they combust readily, releasing vast amounts of heat energy.

Complete combustion

Complete combustion of fuels occurs in a plentiful supply of air and produces carbon dioxide and water. The carbon dioxide released is implicated in the greenhouse effect, which may contribute to climate change.

For example, the equation for the complete combustion of ethane would be:

$$C_2H_6 + 3\frac{1}{2}O_2 \rightarrow 2CO_2 + 3H_2O$$

or

$$2C_2H_6 + 7O_2 \rightarrow 4CO_2 + 6H_2O$$

Incomplete combustion

Incomplete combustion occurs when there is a limited supply of air. The reaction produces carbon monoxide, water and sometimes soot (unburnt carbon).

For example, decane ($C_{10}H_{22}$) burns in a limited supply of oxygen to produce carbon monoxide and water. The equation for this reaction would be:

$$C_{10}H_{22} + 10\frac{1}{2}O_2 \rightarrow 10CO + 11H_2O$$

Pollutants and the internal combustion engine

The internal combustion engine found in most cars uses alkane fuels. This engine produces a number of pollutants, which are shown in Table 13.

Pollutant	How the pollutant forms	Type of pollution
Carbon monoxide (CO)	Incomplete combustion of the fuel in limited oxygen, e.g. $C_8H_{18} + 8\frac{1}{2}O_2 \rightarrow 8CO + 9H_2O$	CO is a toxic gas
Hydrocarbons	Not all of the fuel burns, and some unburnt hydrocarbons leave the exhaust	Hydrocarbons react with oxides of nitrogen (NO_x) to form ground level ozone (O_3), which is a component of smog. It irritates the eyes and causes respiratory problems

Exam tip

Always read the question carefully. It will provide guidance on the products. For example, if you are asked to write an equation for the reaction of Y in a limited supply of air to produce a solid and water only, you need to realise that the products are carbon (the solid) and water.

Pollutant	How the pollutant forms	Type of pollution
Oxides of nitrogen (NO_x)	Nitrogen from the air reacts with oxygen at high temperature and pressure in the engine $N_2 + O_2 \rightarrow 2NO$ $N_2 + 2O_2 \rightarrow 2NO_2$	Oxides of nitrogen react with unburnt hydrocarbons to produce photochemical smog Oxides of nitrogen dissolve in water to form acid rain
Carbon	Incomplete combustion of the fuel in very limited oxygen	Carbon particles exacerbate asthma

Table 13

The gaseous pollutants can be removed by **catalytic converters** attached to exhausts. A catalytic converter has a honeycomb ceramic structure that is coated with a metal catalyst such as platinum or rhodium. The use of a coating means that less metal is used, which keeps the cost down, and the honeycomb structure provides a large surface area for the reactions to take place, which ensures a faster and more complete reaction.

The conversion of polluting and harmful emissions from car exhausts to less polluting or harmful products is achieved by reactions on the surface of catalytic converters. These conversions include:

- Carbon monoxide and nitrogen monoxide reacting to produce nitrogen and carbon dioxide, which are less harmful products:

$$CO + NO \rightarrow CO_2 + \tfrac{1}{2}N_2$$

- Unburnt hydrocarbons reacting with nitrogen monoxide to produce nitrogen, carbon dioxide and water. For example:

$$C_8H_{18} + 25NO \rightarrow 8CO_2 + 9H_2O + 12\tfrac{1}{2}N_2$$

- Oxides of nitrogen (NO_x: NO or NO_2) being reduced to N_2 and O_2.

$$2NO \rightarrow N_2 + O_2$$

$$2NO_2 \rightarrow N_2 + 2O_2$$

Combustion of sulfur-containing hydrocarbons

Some hydrocarbon fuels contain sulfur as an impurity — the sulfur burns to produce the toxic gas, sulfur dioxide. Sulfur dioxide can cause respiratory problems and can react with water and oxygen in the atmosphere to produce acid rain, which causes a huge environmental problem. Acid rain pollutes the environment by killing vegetation, corroding buildings and killing fish in lakes and rivers.

You need to be able to write equations for the combustion of fuels containing sulfur — the reactions produce carbon dioxide, sulfur dioxide and water.

For example, you may be told that a fuel is contaminated with CH_3SH, and asked to write an equation to show the complete combustion of CH_3SH in air. The equation would be:

$$CH_3SH + 3O_2 \rightarrow CO_2 + 2H_2O + SO_2$$

Knowledge check 22

In cars fitted with catalytic converters, unburned octane reacts with nitrogen monoxide to form carbon dioxide, water and nitrogen only. Write an equation for this reaction and identify the catalyst in the converter.

Sulfur dioxide is an acidic gas, and it can be removed from power station flue gases using calcium oxide (a basic oxide) or calcium carbonate in a **neutralisation** reaction.

$$CaO + SO_2 \rightarrow CaSO_3$$

$$CaCO_3 + SO_2 \rightarrow CaSO_3 + CO_2$$

The calcium sulfate(IV), $CaSO_3$, which is produced is oxidised to calcium sulfate(VI), $CaSO_4$, which is used as a construction material.

$$CaSO_3 + [O] \rightarrow CaSO_4$$

Chlorination of alkanes

Alkanes react with halogens (chlorine and bromine) in the presence of ultraviolet light. A **substitution reaction** occurs. These reactions produce a mixture of halogenoalkanes that have varying numbers of halogen atoms. The reaction of methane with chlorine produces hydrogen chloride and a mixture of chloromethane (CH_3Cl), dichloromethane (CH_2Cl_2), trichloromethane ($CHCl_3$) and tetrachloromethane (CCl_4). Hydrogen atoms are replaced by chlorine atoms.

Some of the reactions are:

$$CH_4 + Cl_2 \xrightarrow{\text{UV light}} CH_3Cl + HCl$$

$$CH_4 + 2Cl_2 \xrightarrow{\text{UV light}} CH_2Cl_2 + 2HCl$$

$$CH_4 + 3Cl_2 \xrightarrow{\text{UV light}} CHCl_3 + 3HCl$$

$$CH_4 + 4Cl_2 \xrightarrow{\text{UV light}} CCl_4 + 4HCl$$

Mechanism for the reaction between chlorine and methane

A mechanism is a detailed step-by-step sequence illustrating how an overall chemical reaction occurs. The reaction between chlorine and methane occurs in three steps.

Step 1: **initiation**

- The pattern for all initiation reactions is:

  molecule $\rightarrow$ two radicals

- In this reaction, the ultraviolet light provides energy to break the Cl–Cl bond. The bond splits equally (**homolytic fission**) and each atom obtains one of the electrons from the bond:

  $$Cl_2 \xrightarrow{\text{UV light}} 2Cl\bullet$$

- Two radicals are produced. A free **radical** is a reactive species with an unpaired electron. The unpaired electron is represented by a dot ($\bullet$), so, for example, Cl$\bullet$ is a chlorine radical. A radical is highly reactive due to the unpaired electron.

Knowledge check 23

Name a toxic gas present in flue gases and suggest how it can be removed.

A **substitution reaction** is one in which one group or atom is replaced by another group or atom.

Exam tip

You must be able to apply your knowledge of these equations. For example, you could be asked to write an overall equation for the reaction between bromine and propane to produce C_3Br_8. Simply replace all the hydrogens on propane with bromine atoms.

$$C_3H_8 + 8Br_2 \rightarrow C_3Br_8 + 8HBr$$

Homolytic fission is the breaking of a covalent bond in a molecule to form identical species as one electron from the bond goes to each atom.

Content Guidance

Step 2: **propagation**

- The pattern for all propagation reactions is:

molecule + radical → radical + molecule

- Here, the equations are:

$CH_4 + Cl\bullet \rightarrow CH_3\bullet + HCl$

$CH_3\bullet + Cl_2 \rightarrow Cl\bullet + CH_3Cl$

($CH_3\bullet$ = methyl radical)

- The propagation steps result in a chain reaction that continues until the radicals are removed in a termination step.

Step 3: **termination**

- The pattern for all termination reactions is:

2 radicals → molecule

- Here, the equations are:

$CH_3\bullet + Cl\bullet \rightarrow CH_3Cl$

$CH_3\bullet + CH_3\bullet \rightarrow C_2H_6$

$Cl\bullet + Cl\bullet \rightarrow Cl_2$

- The Cl_2 can then be broken down by ultraviolet light again, so the reaction may not finish.

Reaction mechanisms of this type are called **free-radical substitution reactions**.

Knowledge check 24

What is a radical? Write the formula for a bromine radical.

Exam tip

You must be able to apply this mechanism to the reaction of any alkane with any halogen.

Worked example

Write equations for the steps in the mechanism for the reaction of bromine with propane to form 1-bromopropane.

Answer

Initiation:

$Br_2 \rightarrow 2Br\bullet$

Propagation:

$Br\bullet + CH_3CH_2CH_3 \rightarrow CH_3CH_2CH_2\bullet + HBr$

$CH_3CH_2CH_2\bullet + Br_2 \rightarrow CH_3CH_2CH_2Br + Br\bullet$

Termination:

$CH_3CH_2CH_2\bullet + CH_3CH_2CH_2\bullet \rightarrow CH_3CH_2CH_2CH_2CH_2CH_3$

$CH_3CH_2CH_2\bullet + Br\bullet \rightarrow CH_3CH_2CH_2Br$

Exam tip

Be careful with the position of the radical dot — it must be to the side of or above the last carbon, $CH_3CH_2CH_2\bullet$. If it is beside the CH_3 or in the middle of the chain, it is incorrect.

Summary

- Alkanes have the general formula C_nH_{2n+2} and are saturated hydrocarbons.
- Alkanes are obtained from petroleum by fractional distillation. Longer-chain fractions are broken down into shorter ones by cracking. This creates products that are in greater demand and that have a higher value.
- Cracking is the breaking down of large alkanes into smaller molecules by breaking the C–C bonds. It requires high temperatures to break the strong C–C bond. Thermal cracking and catalytic cracking require different conditions and generate different products.
- Alkanes undergo complete combustion in oxygen to produce carbon dioxide and water, and combust in limited oxygen to produce carbon, carbon dioxide and water. If a fuel contains a sulfur impurity, sulfur dioxide can be produced, which causes acid rain. Sulfur dioxide is removed from flue gases by neutralisation with calcium oxide or calcium carbonate.
- The internal combustion engine produces the pollutants NO_x, CO, unburnt hydrocarbons and carbon. The gaseous pollutants can be removed using a catalytic converter, which has a honeycomb structure to provide a large surface area, and a platinum catalyst.
- Alkanes react with chlorine in the presence of ultraviolet light by a free-radical substitution mechanism.

Halogenoalkanes

Nucleophilic substitution

Halogenoalkanes have the general formula $C_nH_{2n+1}X$, where X is the halogen atom. The carbon–halogen bond (C–X) in halogenoalkanes is polar. The $\delta+$ carbon is electron deficient and can be attacked by nucleophiles, for example, OH^-, CN^-, NH_3. The **nucleophile** donates its electrons. There are three nucleophilic substitution reactions that you need to know, together with their mechanisms.

Nucleophilic substitution of halogenoalkanes by hydroxide ion

General equation: halogenoalkane + $OH^- \rightarrow$ alcohol + X^- (halide ion)

For example:

$CH_3CH_2Br + NaOH \rightarrow CH_3CH_2OH + NaBr$

$CH_3CH_2Br + OH^- \rightarrow CH_3CH_2OH + Br^-$
1-bromoethane   ethanol

This reaction is not used to make ethanol in an industrial setting because the yield is low and bromoethane has to be made first, which is expensive.

- *Conditions*: halogenoalkane is dissolved in a little ethanol; heat under reflux (Figure 33) with aqueous sodium hydroxide

A **nucleophile** is a lone pair donor. It is an atom or group that is attracted to an electron-deficient centre, where it donates the lone pair to form a new covalent bond.

Knowledge check 25

Explain why the carbon–halogen bond in halogenoalkanes is polar.

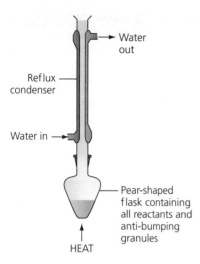

Figure 33

Organic reactions are slow and need heat, but most organic compounds are volatile, and heating without a condenser would cause them to boil off. A reflux condenser (Figure 33) is a condenser that is fitted vertically above a flask. The condenser must be open at the top. Heating under reflux means the continuous boiling and condensing of a reaction mixture. Any vapour formed that escapes from the liquid reaction mixture is changed back into liquid and returned to the liquid mixture.

■ *Mechanism*: nucleophilic substitution (Figure 34)

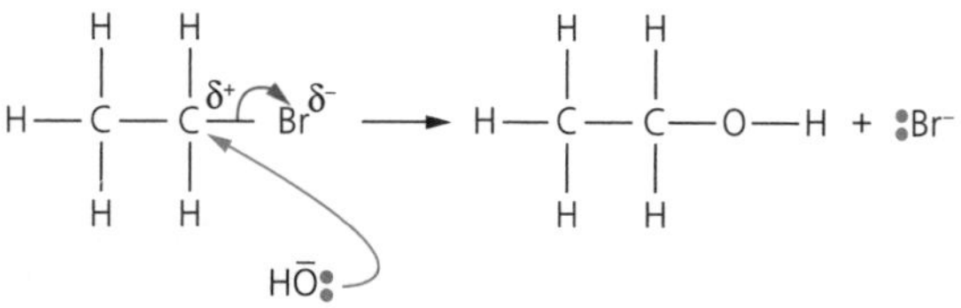

Figure 34

Exam tip

Be careful when drawing curly arrows in a mechanism. Curly arrows represent the movement of electron pairs. The formation of a covalent bond is shown by a curly arrow that starts from a lone electron pair or from another covalent bond. Similarly, the breaking of a covalent bond is shown by a curly arrow starting from the bond.

This reaction can also be classified as a **hydrolysis** reaction — a reaction in which bonds (C–X) are broken by water molecules. Hydrolysis can be catalysed by alkali, as in the bromoethane example.

Rate of reaction of hydrolysis of halogenoalkanes

When halogenoalkanes react, it is the carbon–halogen bond (C–X) that breaks. Going down the group, the bonds get weaker, and have lower **bond enthalpy**, as the shared electrons are further away from the nucleus. As a result the ease of hydrolysis of halogenoalkanes increases down the group — iodo compounds, which have the weakest bonds of lowest bond enthalpy, are hydrolysed most rapidly. The bond enthalpies of the C–X bond are given in Table 14.

Ease of hydrolysis can be tested experimentally by placing equal amounts of 1-chlorobutane, 1-bromobutane and 1-iodobutane in three separate test tubes, each containing $2\,cm^3$ of ethanol and $1\,cm^3$ of silver nitrate solution. On warming in a water bath, a yellow precipitate is observed in the third test tube first, followed by a cream precipitate in the second test tube. After a significant time, a white precipitate usually forms in the first test tube.

Nucleophilic substitution of halogenoalkanes by cyanide ion

General equation: halogenoalkane + CN^- → nitrile + X^- (halide ion)

For example:

$$CH_3CH_2Br + CN^- \rightarrow CH_3CH_2CN + Br^-$$
bromoethane   propanenitrile

This reaction lengthens the carbon chain. Bromoethane (two carbons) is converted to propanenitrile (three carbons).

- *Conditions:* halogenoalkane dissolved in ethanol; heat under reflux with aqueous solution of sodium cyanide or potassium cyanide (NaCN or KCN)
- *Mechanism:* nucleophilic substitution (Figure 35)

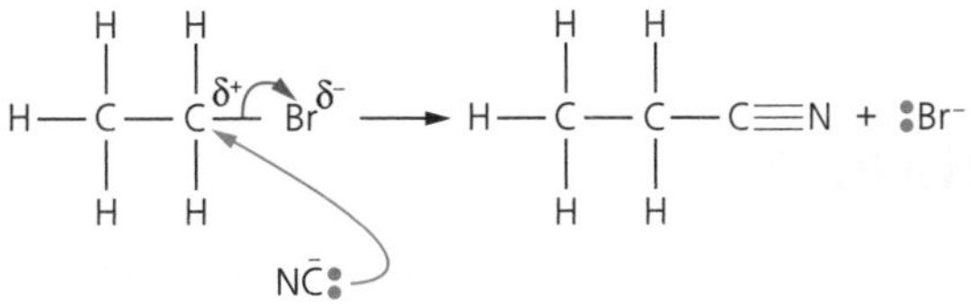

Figure 35

Bond	Bond enthalpy/ $kJ\,mol^{-1}$
C–F	484
C–Cl	327
C–Br	209
C–I	200

Table 14

The **bond enthalpy** of C–X is the energy required (at constant pressure) to break the carbon–halogen bond.

Knowledge check 26

Will 1-bromobutane or 1-chlorobutane react faster with sodium hydroxide? Explain your answer.

Exam tip

When you write mechanisms involving nucleophiles you must show the lone pair. In this example you must show both the lone pair and the negative sign on the cyanide ion.

Exam questions often ask you to outline a mechanism. For example, if you are asked to outline the mechanism of the reaction of bromoethane and cyanide ions, you simply need to draw Figure 35 to answer the question. No words of explanation are necessary.

Nucleophilic substitution of halogenoalkanes by ammonia

General equation: halogenoalkane + $2NH_3 \rightarrow$ amine + ammonium halide

For example:

$CH_3CH_2Cl + 2NH_3 \rightarrow CH_3CH_2NH_2 + NH_4Cl$
chloroethane   ethylamine

- *Conditions*: halogenoalkane dissolved in ethanol; heat with excess concentrated ammonia solution in a sealed tube
- *Mechanism*: nucleophilic substitution (Figure 36)

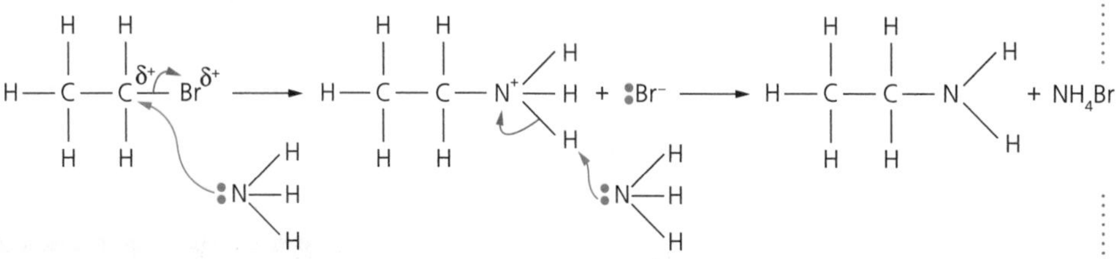

Knowledge check 27

Write an equation for the reaction of bromopropane with ammonia and name the product.

Figure 36

The ammonia has a lone pair of electrons and acts as a nucleophile that attacks the $C^{\delta+}$ of the polar C–X bond. The NH_2 group substitutes for the halogen atom on the halogenoalkane. The amine produced also has a lone pair of electrons on the nitrogen atom so it can act as a nucleophile. This can attack another molecule of chloroethane, causing further substitution and so continuing the reaction.

$CH_3CH_2Cl + CH_3CH_2NH_2 \rightarrow (CH_3CH_2)_2NH + HCl$

If excess ammonia is used then a primary amine is the major product. If excess halogenoalkane is used then successive substitution is more likely to occur and the reaction produces ethylamine, diethylamine and triethylamine.

Elimination reactions of halogenoalkanes

The reagent sodium hydroxide has a role as both a nucleophile and a base. The hydroxide ions react with halogenoalkanes to form an alcohol in a nucleophilic substitution reaction. However, when dissolved in ethanol, the hydroxide ions act as a base and accept a hydrogen ion to form water. As a result, the halogenoalkane molecule loses a hydrogen atom and a halogen atom, to form an alkene. This is an **elimination reaction**.

An **elimination reaction** is one in which a small molecule such as a hydrogen halide is removed from a larger molecule.

Reaction of halogenoalkanes with potassium hydroxide (KOH) in ethanol

General equation: halogenoalkane + KOH → alkene + potassium halide + H_2O

For example:

$CH_3CH_2CH_2Br + KOH \rightarrow C_3H_6 + KBr + H_2O$
1-bromopropane   propene

- *Conditions*: halogenoalkane dissolved in ethanol; hot ethanolic potassium (or sodium) hydroxide solution
- *Mechanism*: elimination (Figure 37)

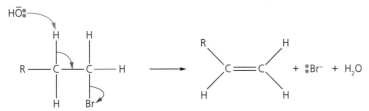

Figure 37

Formation of isomers in the elimination product mixture

In Figure 38, the stars indicate which of the hydrogens might bond to the OH⁻ ion during an elimination reaction. Depending on which of the hydrogens do bond, there are two different structures that may be produced — they are isomers of each other.

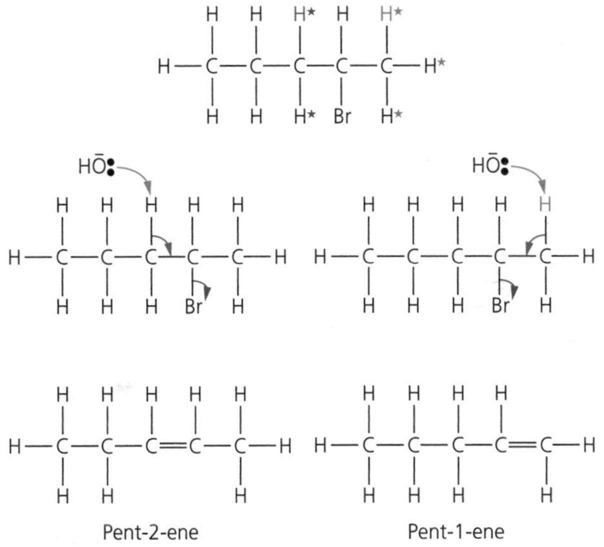

Pent-2-ene   Pent-1-ene

Figure 38

In practice, elimination and substitution reactions both occur simultaneously. The reaction that predominates depends on the reaction conditions and the type of halogenoalkane.

- Hydroxide ions dissolved in water favour substitution, and hydroxide ions dissolved in ethanol favour elimination.
- Primary halogenoalkanes favour substitution, and tertiary halogenoalkanes favour elimination.

Knowledge check 29

Name the organic product formed when bromobutane reacts with:

a sodium hydroxide in ethanolic solution

b sodium hydroxide in aqueous solution

Knowledge check 28

Write an equation for the reaction of 1-bromobutane with potassium hydroxide in ethanol.

Exam tip

The simple way to work out the type of halogenoalkane is to count the alkyl groups bonded to the carbon atom that has the halogen bonded to it. If this carbon atom has one alkyl group bonded to it, then the halogenoalkane is a primary halogenoalkane. If there are three alkyl groups it is a tertiary halogenoalkane.

Ozone depletion

Ozone (O_3) is found in the upper atmosphere in small amounts. It is formed naturally when an oxygen molecule is broken into two radicals by ultraviolet light.

$$O_2 \rightarrow 2O\bullet$$

The oxygen radicals react with oxygen to form ozone:

$$O_2 + O\bullet \rightarrow O_3$$

The presence of ozone in the upper atmosphere is beneficial as it acts like a shield, absorbing ultraviolet radiation, which can cause sunburn, skin cancer and ageing of the skin.

Chlorofluorocarbons (CFCs) cause thinning of the protective ozone layer. This thinning is commonly referred to as holes in the ozone layer. This is because the carbon–chlorine bonds in CFCs are broken down by ultraviolet light, to produce chlorine radicals.

For example:

$$CCl_3F \rightarrow CCl_2F\bullet + Cl\bullet$$

These chlorine radicals react with ozone molecules:

$$Cl\bullet + O_3 \rightarrow O_2 + ClO\bullet$$

$$ClO\bullet + O_3 \rightarrow 2O_2 + Cl\bullet$$

The chlorine radical is regenerated and it can attack more ozone in a chain reaction.

Adding these two equations gives:

$$2O_3 \rightarrow 3O_2$$

The chlorine radical is not destroyed, and so it acts as a catalyst in the decomposition of ozone.

CFCs were used extensively as solvents and refrigerants, as they were unreactive, non-flammable and non-toxic. However, research by many scientists produced evidence of the damage that CFCs cause to the ozone layer, and so legislation was brought in to ban the use of CFCs. Chemists have developed alternative, chlorine-free compounds to replace CFCs and other compounds. One example is trifluoromethane, which cannot produce chlorine radicals and so causes less harm to the ozone layer.

Knowledge check 30

Why are CFCs causing holes in the ozone layer?

Summary

Reagent	Conditions	Mechanism	Product
Sodium hydroxide	Halogenoalkane is dissolved in a little ethanol Heat under reflux with aqueous sodium hydroxide	Nucleophilic substitution	Alcohol
Potassium cyanide	Halogenoalkane dissolved in ethanol Heat under reflux with aqueous solution of KCN (or NaCN)	Nucleophilic substitution	Nitrile
Concentrated ammonia	Halogenoalkane dissolved in ethanol Heated with excess concentrated ammonia solution in a sealed tube	Nucleophilic substitution	Amine
Sodium hydroxide	Halogenoalkane dissolved in ethanol Hot ethanolic potassium (or sodium) hydroxide solution	Elimination	Alkene

- Ozone in the upper atmosphere is beneficial as it absorbs ultraviolet radiation that can cause sunburn, skin cancer and ageing of the skin. CFCs cause thinning or holes in the ozone layer, as the carbon–chlorine bonds are broken down by ultraviolet light to produce chlorine radicals that act as catalysts in the decomposition of ozone.

Alkenes

Structure, bonding and reactivity

Alkenes are **unsaturated hydrocarbons** with the general formula C_nH_{2n}.

A **diene** is an alkene with two C=C bonds in the main carbon chain. Examples include:

$CH_2=C=CH–CH_3$ buta-1,2-diene

$CH_2=CH–CH=CH_2$ buta-1,3-diene

$CH_2=CH–CH_2–CH=CH_2$ penta-1,4-diene

Alkenes contain a double C=C covalent bond. The C=C bond in alkenes is a centre of high electron density, which makes it open to attack by electron-deficient species (electrophiles). This explains why alkenes are much more reactive than alkanes. Alkenes undergo addition reactions — it is said that addition occurs across the double bond.

Addition reactions of alkenes

The C=C bond is a centre of high electron density and is readily attacked by **electrophiles**. As alkenes are unsaturated they can undergo **addition reactions**. Most of the reactions of alkenes are electrophilic addition reactions.

Addition of HBr to alkenes

General equation: alkene + HBr → bromoalkane

For example:

$CH_2CH_2 + HBr → CH_3CH_2Br$
ethene   bromoethane

A molecule of an **unsaturated hydrocarbon** contains at least one C=C or one C≡C bond.

Knowledge check 31

Give the IUPAC name for $CH_2=CHCHClCH_3$ and explain why it is not a hydrocarbon.

An **electrophile** is an electron pair acceptor that attacks a region of high electron density and can accept a lone pair of electrons.

In an **addition reaction** the double covalent bond is broken and two species add on across the double bond to produce one product.

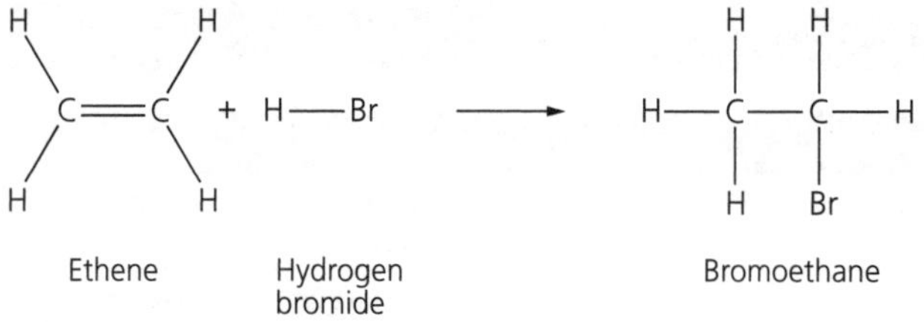

Figure 39

- *Conditions:* bubble in HBr gas at room temperature
- *Mechanism:* electrophilic addition

The mechanism for the reaction of ethene with HBr is shown in Figure 40.

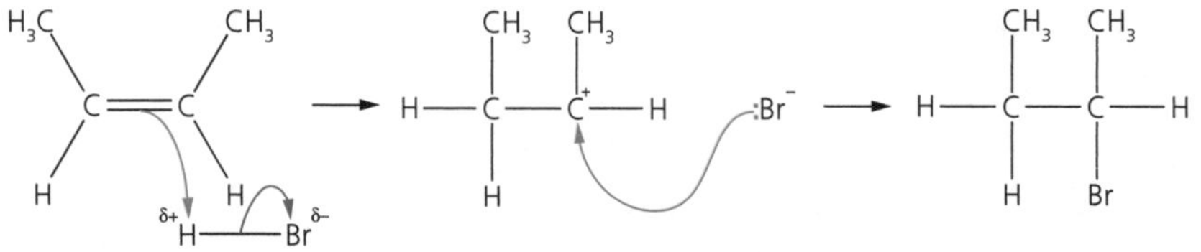

Figure 40

The intermediate formed is called a **carbocation**. The reaction between the carbocation and the bromide ion is rapid as it is caused by the ionic attraction.

The HBr undergoes **heterolytic fission**, as the two species formed are oppositely charged ions.

Addition of H_2SO_4 to alkenes

General equation: alkene + $H_2SO_4 \rightarrow$ alkyl hydrogensulfate

For example:

$$CH_2=CH_2 + H_2SO_4 \rightarrow CH_3CH_2(HSO_4)$$
ethene   ethyl hydrogensulfate

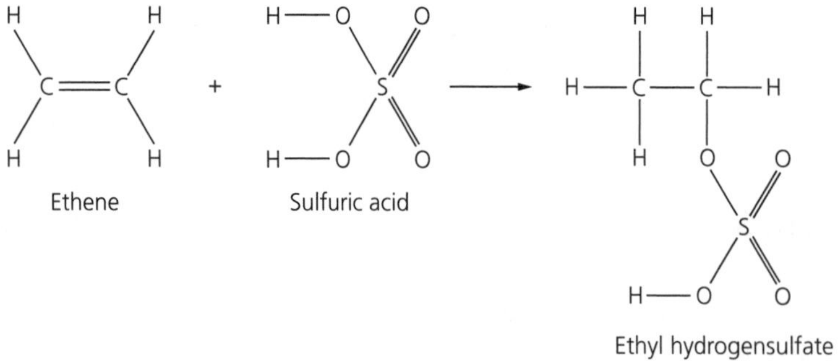

Figure 41

- *Conditions:* use concentrated sulfuric acid at room temperature
- *Mechanism:* electrophilic addition

Exam tip

A carbocation is a positive ion that has the positive charge on a carbon atom. Carbocations are formed as reactive intermediates in some organic reactions, including electrophilic addition reactions.

Heterolytic fission occurs when a covalent bond breaks and both electrons in the bond move to one of the two atoms. Two oppositely charged ions are formed.

Exam tip

The reaction of HBr with any alkene follows a similar mechanism, as does the chlorination of an alkene with HCl.

Knowledge check 32

Name the mechanism by which HBr reacts with but-1-ene.

The mechanism for the reaction of but-2-ene with sulfuric acid is shown in Figure 42.

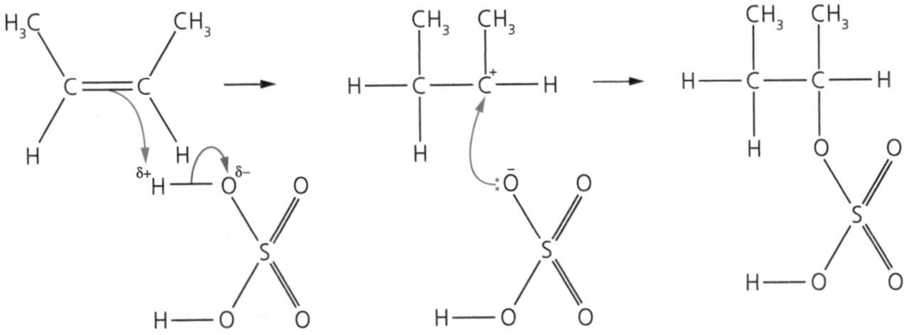

Figure 42

Knowledge check 33

Name the product of the reaction of propene with concentrated sulfuric acid.

Addition of Br$_2$ to alkenes

General equation: alkene + bromine → dibromoalkane

For example:

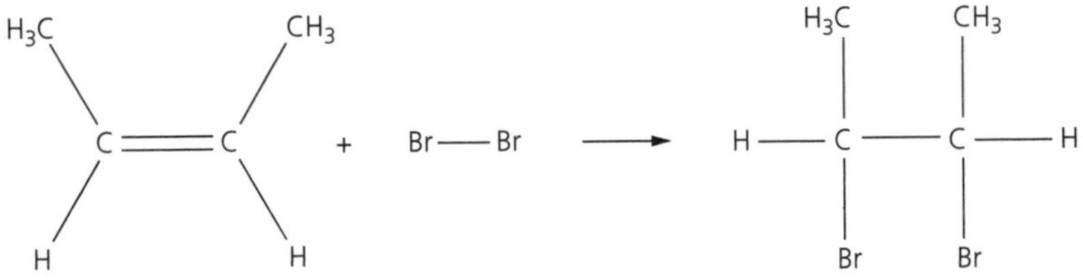

But-2-ene   Bromine   2,3-dibromobutane

Figure 43

- *Conditions:* shake with bromine water at room temperature
- *Mechanism:* electrophilic addition

The mechanism for the reaction between but-2-ene and bromine is shown in Figure 44.

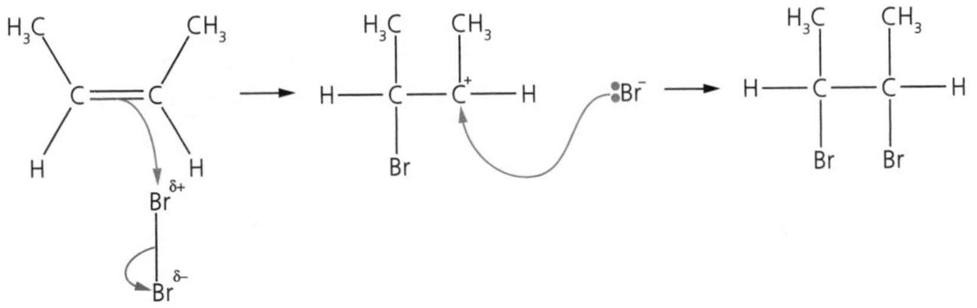

Figure 44

Exam tip

The chlorination of an alkene with Cl$_2$ works similarly. However, chlorine water is virtually colourless so this is not a valid test for unsaturation because it is too difficult to detect the colour change.

Use of bromine to test for unsaturation

An unsaturated compound contains a double bond. Bromine can be used to test for unsaturation as it adds on to the double bond and a colour change occurs. The test is carried out as follows:

- Shake the unknown compound with bromine water.
- If the bromine water changes from orange to colourless then the unknown compound is unsaturated.
- If the bromine water does not change colour then the unknown compound is saturated.

Knowledge check 34

What reagent would you add to distinguish between a sample of cyclohexane and cyclohexene? Describe what you would observe.

Use of excess reagent

Sometimes questions are set on molecules that have multiple functional groups —
for example, two C=C. If an excess of a reagent is used, all of the functional groups
will react. The reaction of penta-1,4-diene with an excess of bromine is shown in
Figure 45. The product is 1,2,4,5-tetrabromopentane. In this example, all of the C=C
react when an excess of a reagent is present.

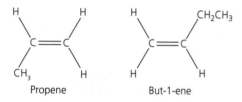

Penta-1,4-diene   1,2,4,5-tetrabromopentane

Figure 45

Addition reactions of unsymmetrical alkenes

An unsymmetrical alkene is one in which the groups or atoms attached to either end
of the C=C are different, for example, propene and but-1-ene are both unsymmetrical,
as shown in Figure 46.

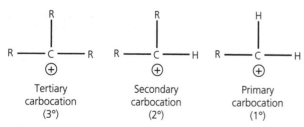

Propene   But-1-ene

Figure 46

When unsymmetrical alkenes undergo addition reactions there may be two different
products, the **major product**, which is present in greater proportion, and the **minor
product**.

The major product is formed via the most stable carbocation intermediate. Alkyl
groups can donate electrons. The more alkyl groups a carbocation has, the more
electrons are donated, decreasing the size of the positive charge on the carbon and
increasing the stability of the carbocation. The order of stability of carbocations is
tertiary > secondary > primary.

Tertiary carbocation (3°)	Secondary carbocation (2°)	Primary carbocation (1°)

where R = alkyl group

Figure 47

> **Exam tip**
>
> Carbocations are
> classified as primary,
> secondary and tertiary,
> depending on the
> number of alkyl groups
> attached. A tertiary
> carbocation has three
> alkyl groups attached,
> as shown in Figure 47.

For example, in the reaction between but-1-ene and hydrogen bromide two different products can form: 1-bromobutane and 2-bromobutane. The major product is 2-bromobutane, as it forms via the more stable secondary carbocation. The minor product is 1-bromobutane as it forms via the less stable primary carbocation (Figure 48).

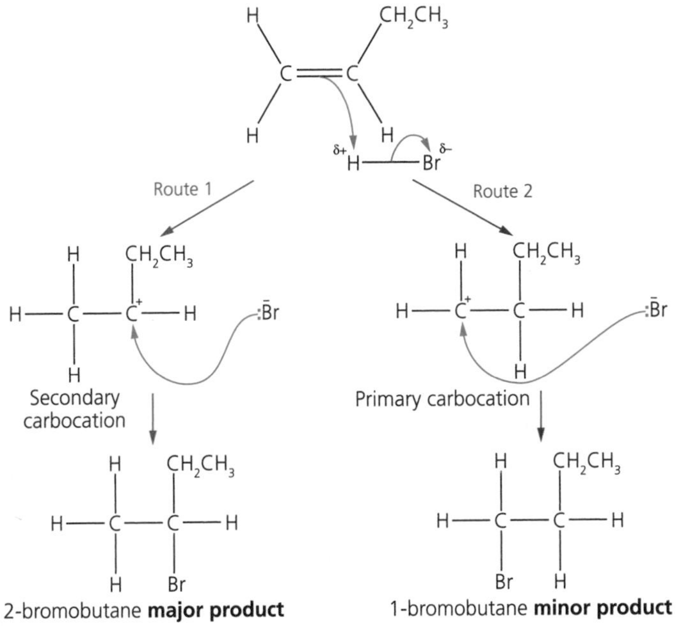

Figure 48

Exam tip

There is an easy way to work out the major product — remember that when hydrogen halides add on to an unsymmetrical alkene the hydrogen becomes attached to the carbon that has the most hydrogen atoms attached.

Knowledge check 35

Draw the structure of the alkene that would react with bromine to produce 1,2-dibromo-3-methylbutane.

Addition polymers

Alkenes react with other alkene molecules to form **addition polymers**, in reactions called addition polymerisation. The polymer names are based on the alkene from which they are formed. For example, ethene forms poly(ethene) — the common name is polythene. The brackets indicate that the structure does not contain the actual alkene but is the polymer formed from the alkene.

Poly(ethene)

Polythene is formed when ethene molecules add on to each other to form a chain of carbon atoms. The unit that breaks its double bond is called the **monomer**, and the long chain formed is called the **polymer**. Polymers no longer have a double bond, and as a result they are unreactive.

An **addition polymer** is a long-chain molecule formed by the repeated addition reactions of many alkene monomers.

A **monomer** is a small molecule that combines with other monomers to make a polymer.

This may be represented as shown in Figure 49:

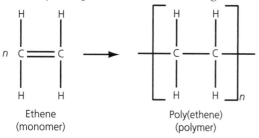

Ethene
(monomer)

Poly(ethene)
(polymer)

Figure 49

The monomer in this addition polymer is ethene and the letter n on both sides of the equation represents a large number of ethene molecules that have to add to each other. The structure inside the brackets is referred to as the **repeating unit** (Figure 49).

Poly(chloroethene)

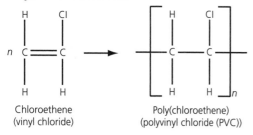

Chloroethene
(vinyl chloride)

Poly(chloroethene)
(polyvinyl chloride (PVC))

Figure 50

Poly(chloroethene) is commonly called PVC, or polyvinylchloride (Figure 50), as the monomer, chloroethene, was commonly called vinyl chloride.

Repeating units

You may often be asked to circle or draw the monomer or the repeating unit from a section of the polymer chain.

Worked example

The structure in Figure 51 shows part of a polymer chain. Draw the monomer and the repeating unit.

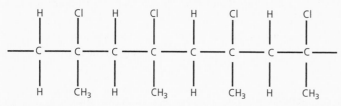

Figure 51

Answer

It is important to be able to identify the repeating unit from the polymer. There are two possibilities, which are circled in Figure 52.

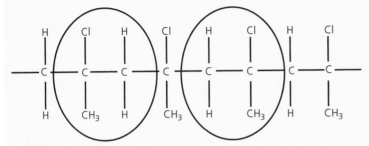

Figure 52

Both of these repeating units would come from the same monomer, as shown in Figure 53.

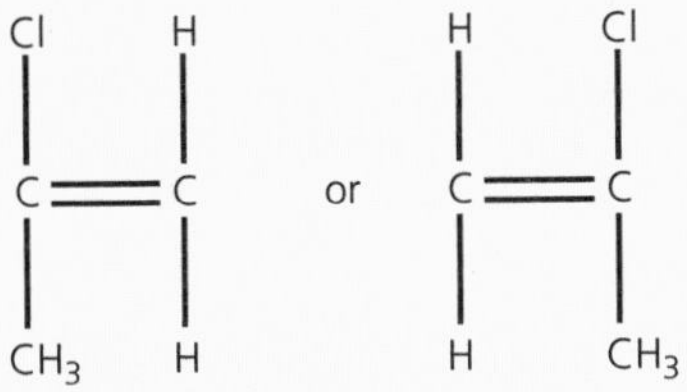

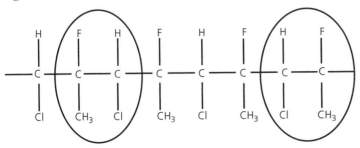

Figure 53

These monomers are both the same — 2-chloropropene. When drawing the monomer make sure it has a double bond.

The repeating unit of this polymer is shown in Figure 54.

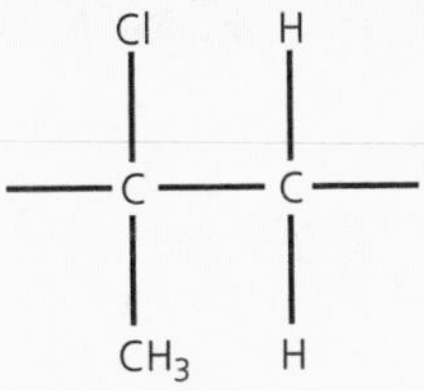

Figure 54

Sometimes the repeating unit can give different monomers. An example is shown in Figure 55.

Figure 55

Exam tip

You should not put brackets around a repeating unit and do not put an *n* after it to indicate the repeat. All bonds need to be shown in a repeating unit, including the bonds to show where the structure would connect to repeating units on either side.

The two possible monomers of this are shown in Figure 56.

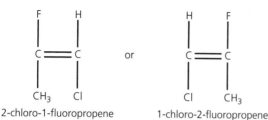

2-chloro-1-fluoropropene   1-chloro-2-fluoropropene

Figure 56

Production and uses of polymers

Natural polymers have been known for many years, but the synthetic polymer industry began only in the 1900s. Today the use of new catalysts has modified the reaction conditions required for the creation of some polymers.

Uses of PVC

The uses of PVC are shown in Table 15. The properties of PVC can be modified using a plasticiser. **Plasticisers** are chemical substances that are used to make a polymer like PVC softer and more flexible. The most common plasticisers used for PVC are phthalates.

	Unplasticised PVC (uPVC)	Plasticised PVC
Property	Hard, rigid and inflexible	Softer and flexible
Use	Window and door frames	Electrical wire insulation, clothing and Wellington boots

Table 15

Summary

- Alkenes are unsaturated hydrocarbons that contain a C=C functional group, which is a centre of high electron density and open to attack by electrophiles.
- The presence of C=C means that alkenes can undergo addition reactions with HBr, Br_2 and H_2SO_4.
- When hydrogen halides add on to an unsymmetrical alkene the hydrogen becomes attached to the carbon that has the most hydrogen atoms attached — this is the major product.
- Alkenes form addition polymers such as poly(ethene) and poly(propene).

Alcohols

Alcohols contain the hydroxyl functional group (–OH). The general formula is $C_nH_{2n+1}OH$. The highly polar O–H bond allows molecules of an alcohol to form hydrogen bonds with other alcohol molecules, and with molecules such as water. Short-chain alcohols are soluble in water. Strong hydrogen bonding between molecules results in a high boiling point, so short-chain alcohols are liquids and higher alcohols are solid.

Alcohol production

Industrial production of ethanol by fermentation

In fermentation, yeast converts sugars such as glucose into ethanol and carbon dioxide.

$$C_6H_{12}O_6(aq) \rightarrow 2CH_3CH_2OH(aq) + 2CO_2(g)$$

- *Conditions:*
  - in the presence of yeast — yeast produces an enzyme that converts glucose to ethanol and carbon dioxide
  - a temperature of 35°C. The enzyme in yeast works best at around this temperature — above this temperature and the enzyme is denatured; below this temperature and the reaction is too slow
  - air is kept out to prevent oxidation of the ethanol formed to ethanoic acid (vinegar)
  - in aqueous solution

Once the fermentation solution reaches about 15% ethanol the yeast can no longer function and fermentation stops.

Ethanol produced industrially by fermentation is separated by fractional distillation and can then be used as **biofuel**.

A **carbon neutral** activity is one that results in no net annual emission of carbon dioxide into the atmosphere. A fuel can be considered carbon neutral when the carbon dioxide produced when the fuel is manufactured and burnt is equal to the amount of carbon dioxide that is used when the raw material is grown. Theoretically, as shown in Table 16, bioethanol can be considered a carbon neutral fuel.

Carbon dioxide removed from the atmosphere	Carbon dioxide released to the atmosphere
In photosynthesis: $6CO_2 + 6H_2O \rightarrow C_6H_{12}O_6 + 6O_2$	In fermentation: $C_6H_{12}O_6 \rightarrow 2C_2H_5OH + 2CO_2$ Four molecules of CO_2 released
	In combustion: $2C_2H_5OH + 6O_2 \rightarrow 4CO_2 + 6H_2O$ (Note that two molecules of ethanol are burnt to use all the ethanol produced in fermentation) Four molecules of CO_2 released
Total: Six molecules of CO_2 removed	Total: Six molecules of CO_2 released

Table 16

Exam tip

Ethanol can be written CH_3CH_2OH or C_2H_5OH in this equation, but not as C_2H_6O or C_2H_5HO.

Knowledge check 37

State three essential conditions for the fermentation of aqueous glucose.

A **biofuel** is a fuel that is made from renewable plant material. Bioethanol is a biofuel.

Knowledge check 38

Write an equation for the complete combustion of butan-1-ol, and state one condition that is required to ensure complete combustion. Why is butan-1-ol described as a biofuel?

Although the ethanol produced by fermentation may in theory be carbon neutral, in practice it may not be totally carbon neutral. This is because:

■ The machinery that is used for planting, harvesting and transporting the sugar crop is powered by burning fossil fuels, which releases carbon dioxide.

■ Fractional distillation of the ethanol to extract it from the fermentation mixture is a process that requires energy from fossil fuels, which releases carbon dioxide.

■ Production of fertilisers for use on the sugar crop may use energy from fossil fuels, which will release carbon dioxide.

Using glucose from crops to manufacture carbon neutral fuel is a major advantage. Another advantage of the use of crops as the raw material for the production of ethanol is that they are sustainable — the crops are a **renewable** resource. There are, however, also disadvantages to the use of crops for the production of ethanol. These include:

■ The depletion of our food supply, as land is increasingly being used to grow crops for fuel. This is an **ethical** issue — in some countries crops that could be used to feed people are being used to provide the raw materials for biofuels instead.

■ Production of crops is subject to the weather and climate.

■ It takes a long time to grow the crops.

■ This route leads to the production of a mixture of water and ethanol, which requires separation and further processing.

Industrial production of alcohol by hydration of alkenes

The **hydration** of ethene is carried out by using steam and a catalyst.

$$CH_2=CH_2 + H_2O(g) \rightleftharpoons C_2H_5OH$$

■ *Conditions:*
  - catalyst of concentrated phosphoric acid absorbed on a solid silica surface
  - 60 atm pressure
  - 600 K temperature
  - excess ethene to give a high yield
  - recycle the unreacted ethene and steam back over the catalyst to increase yield

The ethanol produced in this process is pure, but is manufactured from ethene from crude oil — a finite resource.

■ *Mechanism:* electrophilic addition

The mechanism for the formation of ethanol by the reaction of steam in the presence of a phosphoric acid (H_3PO_4) catalyst is shown in Figure 57, where H^+ represents the phosphoric acid.

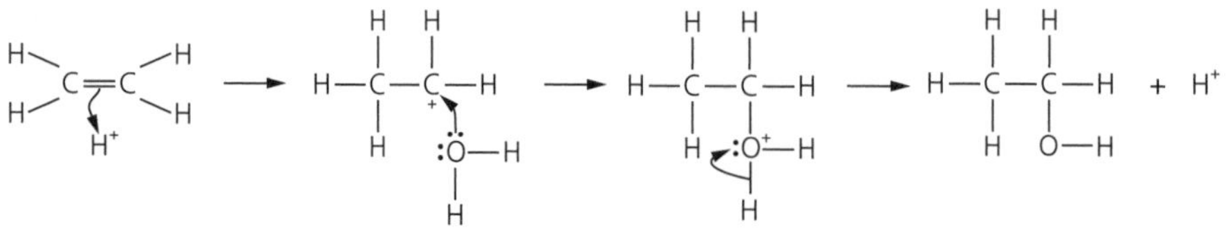

Figure 57

> **Exam tip**
>
> Remember to include equilibrium arrows in this reaction.

> **Hydration** is the addition of water to a molecule.

> **Knowledge check 39**
>
> Use your knowledge of equilibrium reactions to explain why a manufacturer might consider using an excess of steam to react with ethene under the same operating conditions.

Oxidation of alcohols

Classification of alcohols

Alcohols are classified as primary, secondary or tertiary, depending on the number of alkyl groups attached to the carbon bearing the OH group (Figure 58).

- A **primary alcohol** has one alkyl group bonded to the carbon atom that is bonded to the OH group. The exception to this is methanol, which is classed as primary.
- A **secondary alcohol** has two alkyl groups bonded to the carbon that is bonded to the OH group.
- A **tertiary alcohol** has three alkyl groups bonded to the carbon that is bonded to the OH group.

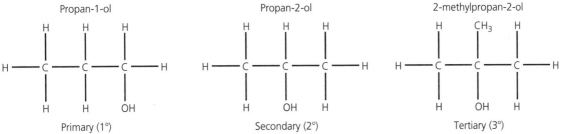

Primary (1°)   Secondary (2°)   Tertiary (3°)

Figure 58

O ...

P... ...tly when heated with acidified
p...

		Method
Pri... oxidi... aldeh... and the carboxy...	...)	To obtain the aldehyde, heat with oxidising agent in distillation apparatus — this removes the aldehyde from the oxidising mixture immediately. To obtain the acid, use excess oxidising agent and heat under reflux
Secondary alcoh... oxidised to ketones only	propan-2-ol propanone $H_3 + H_2O$	Heat under reflux with excess oxidising agent
Tertiary alcohols are not easily oxidised		

Table 17

Knowledge check 40

Explain why butan-1-ol is soluble in water and butane is not.

Exam tip

[O] represents an oxidising agent such as acidified potassium dichromate(VI) solution. It is acceptable to use [O] in equations, as shown in Table 17. If H atoms are removed then the equation is balanced with H_2O on the right-hand side.

Knowledge check 41

Using [O] to represent the oxidising agent, write an equation for the oxidation of butan-1-ol, using acidified potassium dichromate(VI) solution when heating under reflux.

Content Guidance

Required practical 5: Distillation of a product from a reaction

You must be familiar with the apparatus used for distillation, which is shown in Figure 59. You could be asked to draw this apparatus or to describe it. In a description, mention that the mixture is heated in a flask attached to a still head containing a thermometer. A water-cooled condenser is connected to the still head and a collection vessel is placed under it. If you need to collect a liquid with a low boiling point it is important that the collection vessel is cooled in ice to reduce loss by evaporation.

The more volatile liquid (the liquid with the lower boiling point) will evaporate first and the vapour will pass into the condenser, where it condenses on the cool glass and trickles into a collection flask.

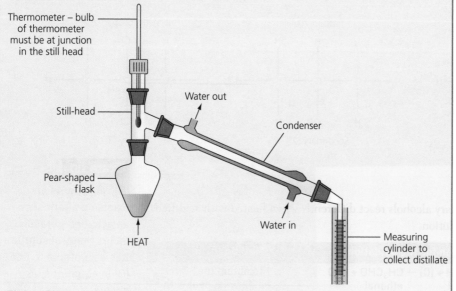

Figure 59

This apparatus can be used to separate ethanal from the reaction mixture of ethanol and acidified potassium dichromate(VI). Some anti-bumping granules should be added to the distillation flask to ensure even boiling. The temperature at which the distillate is collected should be noted. The boiling point of ethanal is low and a cooled collection vessel is necessary to reduce evaporation.

Often, the distillate is impure and can be purified by shaking with an aqueous solution in a separating funnel. When using a separating funnel ensure that you periodically release any pressure that results from gases produced in the reaction, by inverting the funnel and opening the tap. To dry an organic liquid, shake it with a drying agent such as anhydrous sodium sulfate, until it changes from cloudy to colourless, and then filter off the drying agent.

Exam tip

Aldehydes and ketones are further examples of homologous series. In an aldehyde the C=O functional group is on the end of the chain, and it is named '-al'. A ketone is named '-one', and the C=O is in the chain rather than at the end of the chain, as shown:

Propanal

Butanone

Test for aldehydes and ketones

Test	Observation with aldehyde	Observation with ketone	Equation for positive test
Warm with Tollens' reagent	Silver mirror forms on test tube	No reaction — solution remains colourless	$Ag^+ + e^- \rightarrow Ag$ The silver ions are reduced to silver. The aldehyde is oxidised to the acid: $RCHO + [O] \rightarrow RCOOH$
Warm with Fehling's solution	Red precipitate forms	No reaction — solution remains blue	$Cu^{2+} + e^- \rightarrow Cu^+$ Copper(II) ions are reduced to copper(I) oxide, Cu_2O. The aldehyde is oxidised to the acid: $RCHO + [O] \rightarrow RCOOH$

Table 18

Tollens' reagent is a colourless ammoniacal solution of silver(I) nitrate. Fehling's solution is a blue solution that is a mixture of Fehling's solution 1 (copper(II) sulfate solution), and Fehling's solution 2 (a mixture of sodium tartrate and sodium hydroxide solution).

Elimination

Alkenes can be produced from alcohols by acid-catalysed **elimination reactions**. A molecule of water is removed. This reaction can also be termed dehydration (Figure 60).

Propan-1-ol   Propene

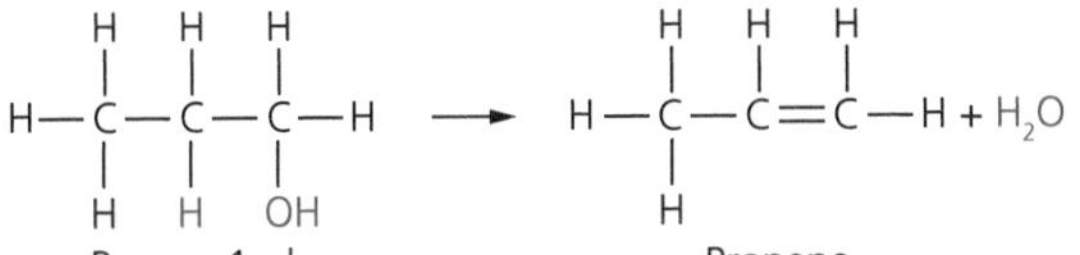

Figure 60

- *Conditions:*
  - concentrated sulfuric/phosphoric acid catalyst
  - temperature of 170°C
- *Mechanism:* elimination

  The mechanism for propan-2-ol is shown in Figure 61.

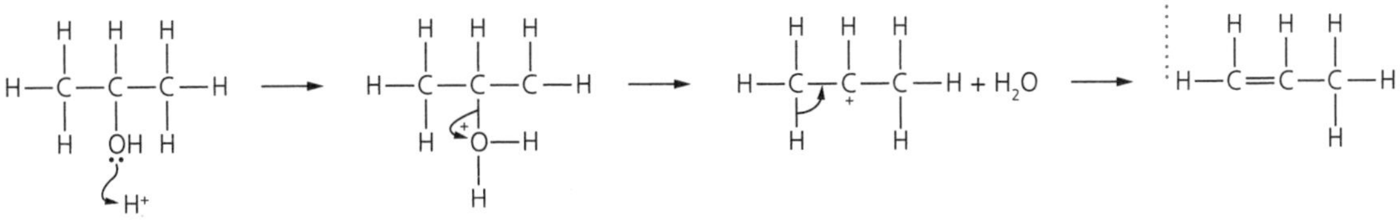

Figure 61

Elimination reactions in an unsymmetrical alcohol

Dehydration of longer-chain alcohols may produce a mixture of alkenes, including E and Z isomers. Figure 62 shows the three products that result from the dehydration of butan-2-ol.

> **Exam tip**
>
> Fehling's solution and Tollens' reagent are mild oxidising agents that can oxidise aldehydes and not ketones, so they can be used to distinguish between samples of the two.

> **Knowledge check 42**
>
> How would you experimentally determine if a substance was propanone and not propanal?

> An **elimination reaction** is one in which a small molecule is removed from a reactant molecule.

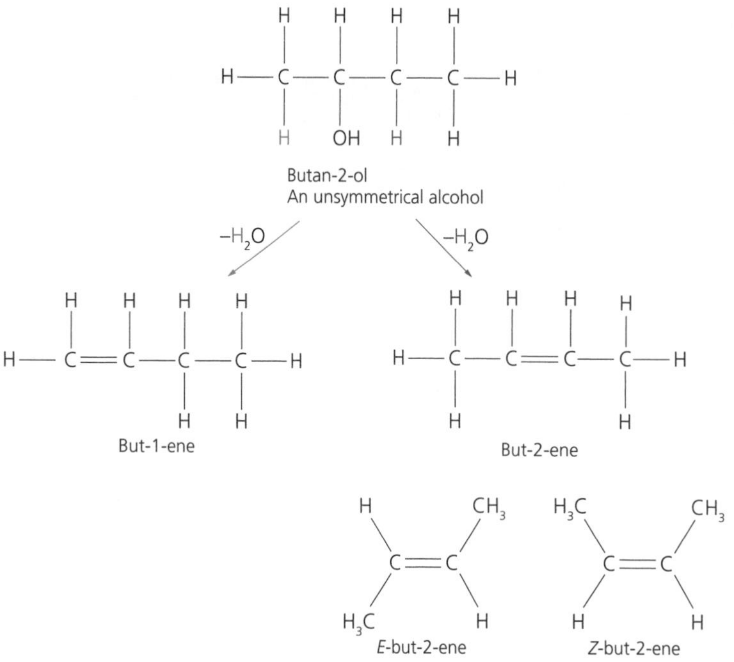

Figure 62

Alkenes produced by this method can be used to generate addition polymers. This is important, as the alcohol used for the dehydration can be produced from renewable crops using fermentation, and so the alkenes produce polymers without using monomers derived from crude oil.

Summary

- Alcohol is produced by the fermentation of sugars using yeast at 35°C in the absence of air.
- A carbon neutral activity is one that results in no net annual emissions of carbon dioxide to the atmosphere. Bioethanol is not totally carbon neutral because the machinery that is used for planting, harvesting and transporting the sugar crop, and the production of fertilisers for the crop, is powered by burning fossil fuels, which releases carbon dioxide.
- Alcohol can also be produced industrially by the hydration of ethene with steam, at 60 atm pressure and 500 K temperature, using a catalyst of phosphoric acid. The mechanism is electrophilic addition.
- Primary and secondary alcohols undergo mild oxidation using acidified potassium dichromate(VI)

solution. Tertiary alcohols do not undergo mild oxidation.
- If the aldehyde is the desired product from the oxidation of a primary alcohol, then the method used is distillation. if a carboxylic acid is desired, then the method used is heating under reflux.
- Aldehydes give a silver mirror when warmed with colourless Tollens' reagent, but a red precipitate forms when an aldehyde is warmed with blue Fehling's solution. Ketones do not react with either reagent.
- Alkenes can be produced from alcohols by acid-catalysed elimination reactions. A molecule of water is removed. A catalyst of concentrated sulfuric or phosphoric acid is used, and a temperature of 170°C.

■Organic analysis

Identification of functional groups by test tube reactions

Table 19 summarises some test tube reactions that can be carried out to identify functional groups. You should be familiar with these tests.

Homologous series	Test	Observation
Alkene	Shake with bromine water	Orange solution turns to colourless solution
Aldehyde	Warm with Tollens' reagent	Colourless solution initially Silver mirror produced on test tube
	Warm with Fehling's solution	Blue solution initially Red precipitate produced
Carboxylic acid	Add some solid sodium carbonate	Solid disappears, bubbles of gas form — gas (CO_2) turns limewater colourless to milky
Primary or secondary alcohol	Warm with acidified potassium dichromate(VI) solution	Orange solution turns to green solution
Tertiary alcohol	Warm with acidified potassium dichromate(VI) solution	Orange solution remains orange

Table 19

> **Required practical 6: Tests for alcohol, aldehyde, alkene and carboxylic acid**
>
> The tests for different homologous series are shown in Table 19. You must be able to carry out test tube reactions for these tests to identify different compounds. Make sure that you are familiar with the observations that would accompany a positive test.

Knowledge check 43

How would you experimentally decide if a sample of an organic liquid was ethanol or 2-methylpropan-2-ol?

Mass spectrometry

Mass spectrometry can be used to determine the molecular formula of a compound. The molecular ion peak corresponds to the relative molecular mass of a molecule. Many organic molecules have the same relative molecular mass — for example, C_6H_{12} and C_5H_8O both have $M_r = 84.0$.

If more precise atomic masses are used then more precise molecular masses are obtained — these are different and so it is possible to determine the molecular formula. For example:

C_6H_{12} has $M_r = 84.093895$

C_5H_8O has $M_r = 84.057511$

Exam tip

Remember that the molecular ion peak is the peak with the largest m/z ratio, and it gives the relative molecular mass of the molecule.

High-resolution mass spectrometers can measure masses to many decimal places, so they allow you to work out a precise relative molecular mass and hence the molecular formula of a compound.

Worked example

A sample of air, when analysed in a mass spectrometer, was found to include traces of a gas that had a molecular ion value of 44.00105. Use the data in Table 20 to show that the trace gas was dinitrogen oxide (N_2O). State why the trace gas could have been propane or carbon dioxide, if relative molecular masses calculated to one decimal place had been used.

	Carbon	Nitrogen	Oxygen
Precise relative atomic mass	12.00000	14.00307	15.99491

Table 20

Answer

The molecular ion value gives the relative molecular mass — it is 44.00105. To show that this is N_2O, simply add together the precise relative atomic mass values of $(2 \times N)$ and $(1 \times O)$.

$2(14.00307) + 15.99491 = 44.00105$

propane is $C_3H_8 = (3 \times 12.0) + (8 \times 1.0) = 44.0$

carbon dioxide is $CO_2 = (1 \times 12.0) + (2 \times 16.0) = 44.0$

dinitrogen oxide is $N_2O = (2 \times 14.0) + (1 \times 16.0) = 44.0$

They have the same M_r (to 1 decimal place) and would give the same molecular ion peak.

Exam tip

To answer the second part of the question you need to calculate the relative atomic mass of each of these molecules using the relative atomic masses, given to 1 decimal place, in your periodic table.

Knowledge check 44

State the m/z value of one peak that you would expect to find in the mass spectrum of ethanoic acid.

Infrared spectroscopy

Bonds between atoms are not rigid — the atoms are free to stretch and bend (often called vibrations). In infrared spectroscopy a beam of infrared radiation is passed through a molecule. The covalent bonds in the molecule absorb the infrared radiation and vibrate more. Any particular bond can only absorb radiation that has the same frequency as the natural frequency of the bond, so different bonds absorb at different characteristic wavenumbers and give peaks (see Figure 63).

The position of the absorption depends on the bond strength and on the masses of the atoms involved.

- strong bonds and light atoms absorb at high wavenumbers
- weak bonds and heavy atoms absorb at low wavenumbers

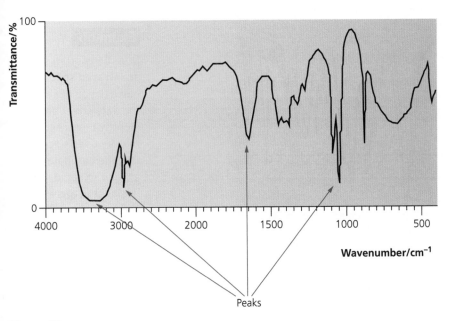

Peaks

Figure 63

The horizontal axis of an infrared spectrum is the wavenumber (measured in cm^{-1}):

$$\text{wavenumber (cm}^{-1}) = \frac{1}{\text{wavenumber (cm)}}$$

The horizontal axis runs from 4000 cm^{-1} on the left to approximately 400 cm^{-1} on the right. The vertical axis is % transmittance. Infrared spectra are plotted as transmittance, so there are dips in the plot from 100% — these dips are known as peaks.

The data sheet that you will be given in your examination contains a table that lists the wavenumbers for some bonds. You are expected to be able to use this data to identify particular bonds and, therefore, functional groups.

An organic reaction can be monitored using infrared spectroscopy. As the reaction proceeds, one functional group is converted into another so the signal for one group decreases while the signal for another group increases. For example, if the oxidation of a secondary alcohol is monitored using infrared spectroscopy, then the absorption peak at 3230–3550 cm^{-1} due to the OH decreases and an absorption at 1680–1750 cm^{-1} appears due to the C=O in the ketone that is produced.

Fingerprint region

The region below 1500 cm^{-1} has many peaks, which are caused by the varied vibrations of entire molecules. It is called the fingerprint region and is unique for each compound, so can be used to identify a chemical. To identify a molecule, an infrared spectrum of a sample is produced and the fingerprint region is compared with those of known compounds in a computer database. An exact match identifies the sample.

Figure 64 shows the infrared spectrum of ethanoic acid superimposed on the infrared spectrum of butanoic acid. They are both similar and have the characteristic C=O and O–H peaks. However, the fingerprint regions are different and are unique to each compound.

Exam tip

The most useful absorptions for identifying organic molecules include:
- a broad O–H peak at 2500–3000 cm^{-1} (carboxylic acid)
- a broad O–H peak at 3230–3550 cm^{-1} (alcohol)
- a sharp, intense C=O peak at 1680–1750 cm^{-1} (aldehydes, ketones, acids or esters)

Exam tip

Remember that just as a fingerprint can be used to identify people, the fingerprint region can be used to identify a chemical by the comparison of spectra.

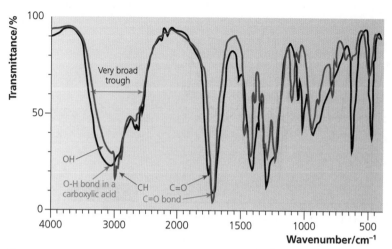

Figure 64 Infrared spectra of ethanoic acid (black) and butanoic acid (blue)

If you are asked to explain how infrared spectroscopy can be used to identify an alcohol:

- State that the first stage is to run the infrared spectrum of the alcohol — it should show an absorption peak in the range 3230–3550 cm^{-1}, which is characteristic of an alcohol.
- State that this spectrum should be compared with database spectra. The spectrum that has the same fingerprint region gives the identity of the alcohol.

Analysis of a spectrum

The infrared spectrum of ethanoic acid is shown in Figure 65.

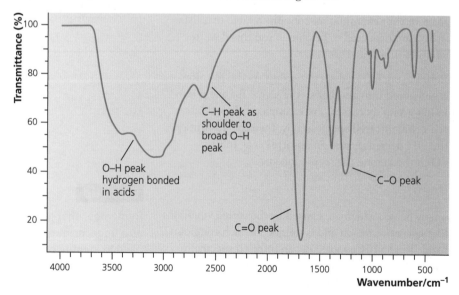

Figure 65

Exam tip

Questions often show an infrared spectrum and ask you to refer to the data sheet and to identify features that may be used to confirm that the compound is an alcohol, acid, etc. It is best to answer by giving wavenumbers and identifying peaks. For example, 'there is an absorption peak at 3230–3550 cm^{-1}, which is characteristic of an OH bond in an alcohol.'

- The broad band absorption between $2500\,cm^{-1}$ and $3700\,cm^{-1}$ suggests the presence of an O–H group that is hydrogen bonded in a carboxylic acid.
- The sharp intense peak at around $1700\,cm^{-1}$ indicates the presence of a C=O bond, which adds to the evidence for a carboxylic acid.
- There is a C–O peak at around $1250\,cm^{-1}$.
- The C–H peak at around $2650\,cm^{-1}$ is caused by C–H bonds.

Conclusion: the spectrum is for a carboxylic acid.

Identifying impurities

Infrared spectra can be used to check the purity of a compound. An infrared spectrum of the compound is produced and is compared with the infrared spectrum of the pure compound. If the compound is impure, there are extra peaks in the spectrum.

Absorption of infrared radiation and global warming

The greenhouse gases carbon dioxide, methane and water vapour have bonds that are good at absorbing infrared radiation emitted from the sun. Oxygen and nitrogen in the atmosphere do not have this property. As a result, if the amount of greenhouse gases in the atmosphere increases, then more infrared radiation is absorbed, leading to the greenhouse effect and global warming.

Knowledge check 45

State one feature on the infrared spectrum of ethanal that would confirm that it is an aldehyde.

Summary

- The functional group in an organic molecule can be identified in various test tube reactions, for example, adding bromine water to identify an alkene.
- High-resolution mass spectrometry records m/z values to many decimal places. Precise molecular masses can be used to determine the molecular formula of a compound.
- Organic molecules absorb infrared radiation due to molecular vibrations (bending and stretching) of covalent bonds.

- An infrared spectrum shows percentage transmittance against wavenumber (cm^{-1}). Bonds in a molecule absorb infrared radiation at characteristic wavenumbers.
- The region below $1500\,cm^{-1}$ is called the fingerprint region and can identify a molecule when compared with known spectra.
- By comparing the infrared spectrum of a substance with that of a pure sample, it is possible to identify impurities in the sample.
- Greenhouse gases in the atmosphere have bonds that can absorb infrared radiation, and this contributes to global warming.

Questions & Answers

This section contains a mix of both multiple-choice and structured questions that are similar to those you can expect to find in the AS and A-level papers 1 and 2.

Papers 1 and 2 of the AS examinations consist of 15 multiple-choice questions (each with four options, A to D), followed by 65 marks of structured questions, totalling 80 marks. The papers are both 1½ hours. Papers 1 and 2 of the A-level consist of short- and long-answer questions (some of which are on the topics covered in this book), totalling 105 marks. They are 2 hours long. Paper 3 of the A-level is a 2-hour paper with 90 marks. It contains 40 marks of questions on practical techniques and data analysis, and 50 marks of questions testing across the complete specification, of which 30 marks are multiple choice. For both the AS and A-level papers, 15% of the marks cover practical aspects of the specification, and 20% mathematical content.

About this section

Each question in this section is followed by brief guidance on how to approach the question and also on where you could make errors (shown by the icon ⊜). The answers given are those that examiners would expect from a top-grade candidate. Answers are followed by comments that explain why the answers are correct and pitfalls to avoid. These are preceded by the icon ⊜. You could try the questions first to see how you get on and then check the answers and comments.

General tips

- Be accurate with your learning at this level as examiners penalise incorrect wording.
- At both AS and A-level, at least 20% of the marks in assessments for chemistry require the use of mathematical skills. For any calculation, always follow it through to the end, even if you feel that you have made a mistake, as there are marks for the correct method even if the final answer is incorrect.
- Always attempt to answer multiple-choice questions (there are 15 marks available on both papers 1 and 2 at AS for multiple-choice questions) even if it is a guess — you have a 25% chance of getting it right.

■ Inorganic chemistry

Periodicity

Question 1

(a) State and explain the general trend in the first ionisation energies of period 2.   (3 marks)

ⓔ You have studied the trend in first ionisation energy across period 3. To answer this question you must simply realise that the trend is exactly the same. Remember that it is not enough to state that the first ionisation energy increases — you must state that it increases across the period.

> **(a)** The first ionisation energy increases across the period. ✓
>
> There is an increase in nuclear charge due to more protons and similar shielding. ✓
>
> Hence, there is a smaller atomic radius and so the outermost electron is held closer to the nucleus by the greater nuclear charge. ✓

(b) State how the element oxygen deviates from the general trend in first ionisation energies across period 3. Explain your answer.   (3 marks)

ⓔ When studying the trends in first ionisation energy you should have noted that the ionisation energies of Group 6 elements are lower than expected. To answer this question it is essential to write the electronic configuration of oxygen ($1s^2 \, 2s^2 \, 2p^4$) and then explain the stability — the paired electrons in the $2p$ orbital repel and less energy is needed to remove one, which decreases the first ionisation energy. Always name the orbital — in this case $2p$.

> **(b)** Ionisation energy of oxygen is lower ✓ because the pair of electrons ✓ in $2p$ repel each other. ✓

(c) A general trend exists in the first ionisation energies of the period 2 elements, lithium to fluorine. Identify one other element, apart from oxygen, which deviates from this general trend.   (1 mark)

> **(c)** Boron ✓

ⓔ You need to remember that the first ionisation energies of group 3 and group 6 are lower than expected. The electronic configuration of boron is $1s^2 \, 2s^2 \, 2p^1$. It has a lower first ionisation energy as the $2p$ electron is further from the nucleus than the stable filled $2s^2$ subshell of beryllium, which is closer to the nucleus.

Group 2

Question 1

Which one of the following describes a correct trend as group 2 is descended
from beryllium to barium?   (1 mark)

A   The first ionisation energy of the metal increases.

B   The reactivity of the metal decreases.

C   The solubility of the hydroxide increases.

D   The solubility of the sulfate increases.

> Answer is C ✓.

ⓔ This question tests your knowledge of trends in group 2. You need to be
familiar with the fact that the reactivity of the metal increases down the group and
that ionisation energy decreases. Try and remember that barium sulfate is used
in X-rays as it is insoluble. Thus, the trend is that the solubility of group 2 sulfates
decreases down the group. Also remember that hydroxides have the opposite
trend — solubility increases down the group.

Question 2

It is possible to distinguish between $BaCl_2(aq)$ and $MgCl_2(aq)$ by adding an aqueous
reagent to a sample. Give a suitable aqueous reagent that could be added
separately to each compound. Describe what you would observe in each case.   (3 marks)

ⓔ These are both group 2 compounds, You need to think about identification
tests and about the solubility of group 2 compounds — you should remember
that magnesium hydroxide is insoluble but magnesium sulfate is soluble, barium
hydroxide is soluble but barium sulfate is insoluble. Adding any soluble sulfate,
for example, sodium sulfate solution or sulfuric acid, to the compounds will form
magnesium sulfate, which is soluble and produces a colourless solution, and
barium sulfate, which is insoluble and produces a white precipitate. Note that
you must give the full name of the reagent — it is not sufficient to write 'add a
soluble sulfate'. Alternatively, sodium or potassium hydroxide solution could be
added. Magnesium hydroxide is formed as an insoluble white precipitate; barium
hydroxide is soluble and a colourless solution remains.

> ■ Add some aqueous sodium sulfate or sulfuric acid (any named soluble sulfate) ✓
> ■ A white precipitate forms for barium chloride solution ✓
> ■ Whereas the solution remains colourless/no reaction for magnesium chloride
>   solution ✓
> *or*
> ■ Add sodium or potassium hydroxide solution ✓
> ■ The solution remains colourless/no reaction for barium chloride solution ✓
> ■ A white precipitate occurs for magnesium chloride solution ✓

Group 7

Question 1

Consider the following reaction in which bromide ions behave as reducing agents.

$$Cl_2(aq) + 2Br^-(aq) \rightarrow Br_2(aq) + 2Cl^-(aq)$$

(a) In terms of electrons, state the meaning of the term reducing agent. (1 mark)

(b) Write a half-equation for the conversion of chlorine into chloride ions. (1 mark)

(c) Suggest why bromide ions are stronger reducing agents than chloride ions. (2 marks)

ⓔ Do not get mixed up between reduction and a reducing agent — reduction is gain of electrons, but a reducing agent is a species that donates or loses electrons. In the formation of chloride ions, you must remember that one chlorine molecule produces two chloride ions, and as a result, two electrons are gained. The bromide ions have more electron shells than do the chloride ions, so they have a larger ionic radius, which means that the electron to be lost is further from the nucleus and is also shielded more from the nuclear charge. As a result the electron in the bromide ion is attracted less strongly to the nuclear charge and is more easily lost. Make sure you mention the strength of nuclear attraction for the electron.

> **(a)** An electron donor ✓
>
> **(b)** $Cl_2 + 2e \rightarrow 2Cl^-$ ✓
>
> **(c)** The bromide ions have a larger ionic radius. ✓
>
> The electron to be lost is further from the attraction of the nucleus and is held less tightly. ✓

Question 2

When excess chlorine is bubbled into cold, dilute, aqueous alkali, which one of the following lists the main products of the reaction? (1 mark)

A Cl^-, ClO^-, H_2O

B Cl^-, ClO_3^-, H_2O

C Cl^+, ClO^-, $HClO$

D ClO^-, Cl^-

ⓔ This question shows how important it is to learn equations for the reactions of the group 7 elements and compounds, in this case, the reaction of chlorine with cold sodium hydroxide (alkali). Write the symbol equation for the reaction of chlorine with sodium hydroxide and remove the spectator ions, and you will quickly work out the products of the reaction.

> Answer is A ✓

Questions & Answers

Question 3

A solid compound was dissolved in deionised water. Dilute nitric acid was added, followed by silver nitrate solution. A white precipitate was observed, which dissolved when ammonia solution was added.

(a) Name the ion that has been identified during this test. (1 mark)

(b) Write the simplest ionic equation for the formation of the white precipitate. (1 mark)

(c) Explain why dilute nitric acid is added to the solution before silver nitrate solution. (2 marks)

ⓔ Silver nitrate solution is the key to this question as it is used to test for halide (chloride, bromide and iodide) ions. Chloride ion gives a white precipitate. Be careful that you do not write down chlorine ion — this is incorrect. The dilute nitric acid is added to remove any carbonate ions, which would give a white precipitate of silver carbonate, Ag_2CO_3. This would be a false positive test for chloride ions. Learn the observations carefully in terms of the colour of the precipitates and whether they redissolve in ammonia solution.

(a) Chloride ✓

(b) $Ag^+ + Cl^- \rightarrow AgCl$ ✓

(c) Nitric acid reacts with/removes carbonate/hydroxide ions. ✓

Carbonate ions would give a white precipitate/false test for chloride ions. ✓

Question 4

Sodium iodide reacts with concentrated sulfuric acid.

(a) Name two solid products of the reaction and one gaseous product. (3 marks)

(b) Write two balanced symbol equations for the reactions that are occurring. (2 marks)

ⓔ Remember that the equations for the reactions of the halides with concentrated sulfuric acid are similar, with the first equation being common to the chloride, bromide and iodide; the second is common to the bromide and iodide. The last two equations must be recalled for the iodide, but knowing the equations allows you to name all the products as well as stating observations for the reactions. So it is best to answer part (b) first and use the equations to answer part (a). If the equations are not asked for, write them down to remind yourself. You will not lose marks for incorrect equations if they are not asked for.

(a) Solids: iodine ✓, sulfur ✓

Gas is hydrogen sulfide/sulfur dioxide ✓

(b) Any two from ✓✓:

$NaI + H_2SO_4 \rightarrow HI + NaHSO_4$

$2HI + H_2SO_4 \rightarrow I_2 + SO_2 + 2H_2O$

$6HI + H_2SO_4 \rightarrow 3I_2 + S + 4H_2O$

$8HI + H_2SO_4 \rightarrow 4I_2 + H_2S + 4H_2O$

Organic chemistry

Introduction to organic chemistry

Question 1

The correct systematic name for the organic compound with the structure shown in Figure 1 is:

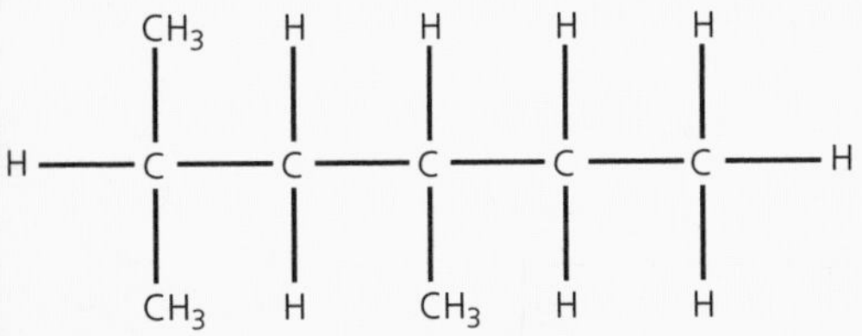

Figure 1

A 1,1,3-trimethylpentane

B 3,5,5-trimethylpentane

C 2,4-dimethylhexane

D 3,5-dimethylhexane

> Answer is C ✓

(e) This type of question tests your ability to count the longest carbon chain and to number it correctly. There can never be a 1-methyl on a substituted alkane as this methyl group is part of the main chain. If there is no functional group to define the numbering, number the carbon chain from the end that gives the lowest locant numbers for any substituent groups.

Question 2

(a) The structure of an organic alcohol (A) is shown in Figure 2. Explain how the Cahn–Ingold–Prelog (CIP) priority rules can be used to deduce the full IUPAC name of this compound.

(6 marks)

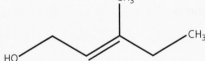

Figure 2

(e) The Cahn–Ingold–Prelog priority rules state that the higher the atomic number of the atom attached directly to the carbon of the C=C, then the higher the priority. Remember that if two atoms are the same, consider the total atomic number of the atoms bonded directly to them. In this case the skeletal formula is given — first, you need to identify the two groups bonded to each carbon of the carbon–carbon double bond; it may help to draw the structural formula (Figure 3).

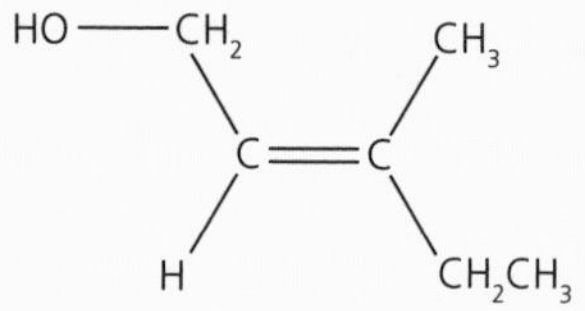

Figure 3

Start with the left-hand carbon of the carbon–carbon double bond, which has a carbon and a hydrogen bonded to it, and decide on the priority based on atomic number. Then move to the right-hand carbon of the carbon–carbon double bond; it has two carbon atoms attached. So, consider the atoms bonded to each of these carbons; three hydrogens bonded to the top carbon and two hydrogens and one carbon bonded to the bottom carbon, which has higher priority.

(a) The Cahn–Ingold–Prelog priority rules state that the higher the atomic number of the atom attached directly to the carbon of the C=C then the higher the priority. ✓

Left-hand carbon:

On the left-hand carbon of the double bond there is one carbon attached (atomic number 6) and one hydrogen (atomic number 1). The higher atomic number takes priority, so the CH_2OH has highest priority. ✓

Right-hand carbon:

On the right-hand carbon of the double bond there are two carbons attached, of the same atomic number and priority. ✓

Moving to the atoms attached to these carbons; on the top carbon in the methyl group there are 3 H (atomic number total of 3). On the bottom carbon of the ethyl group there are 2 H and 1 C directly attached (atomic number total of 8); this group has highest priority. ✓

Conclusion

The highest priority groups (CH_2CH_3) and CH_2OH are on opposite sides of the double bond, so the isomer shown is the *E* isomer. ✓.

The IUPAC name is *E*-3-methylpent-2-en-1-ol. ✓

(b) Give the name of the organic product formed when A is dehydrated. Name the reagent used.

(2 marks)

ⓔ A dehydration reaction removes water from the molecule. This will leave a double bond — the new molecule will have two double bonds and is a diene.

(b) Hexa-1,2-diene ✓; concentrated sulfuric acid ✓

Question 3

How many structural isomers have the molecular formula C_4H_9Br? (1 mark)

A 2

B 3

C 4

D 5

The correct answer is C ✓

ⓔ Structural isomers have the same molecular formula — in this case C_4H_9Br — but different structural formulae. Start by drawing four carbon atoms in a row — there are two structures like this — 1-bromobutane and 2-bromobutane. Remember to name each isomer so that you do not duplicate (there is no 3-bromobutane or 4-bromobutane because these have the same structure as the 1-bromobutane or 2-bromobutane, and would therefore be named 1- or 2- instead). Then draw structures with three carbons in a row and the methyl as a side group — there are two structures: 2-bromo-2-methylpropane, and 1-bromo-2-methylpropane. This gives a total of four structural isomers.

Alkanes

Question 1

Alkanes are used as fuels. A student burned some decane ($C_{10}H_{22}$) in air and found that the combustion was incomplete.

(a) Write an equation for the incomplete combustion of decane to produce carbon monoxide as the only carbon-containing product. Suggest why the combustion was incomplete. (2 marks)

(b) Write an equation for the complete combustion of decane. (1 mark)

(c) Why does this reaction create environmental problems? (1 mark)

(d) One molecule of $C_{10}H_{22}$ can be cracked to form one molecule of pentane and one other product. Write an equation for this cracking reaction. (1 mark)

ⓔ For the incomplete combustion of any alkane, there are three possible products — carbon, carbon monoxide and water. In the question it states that the only carbon-containing product is CO, so you need to write a balanced symbol equation to produce CO and water only. It is best to balance the carbon first, then the hydrogen and finally the oxygen, using half numbers if necessary. Incomplete combustion occurs in limited air.

(a) $C_{10}H_{22} + 10\frac{1}{2}O_2 \rightarrow 10CO + 11H_2O$. ✓ There was limited air. ✓

(b) Complete combustion produces carbon dioxide and water.

$$C_{10}H_{22} + 15\frac{1}{2}O_2 \rightarrow 10CO_2 + 11H_2O ✓$$

(c) Carbon dioxide causes the greenhouse effect and contributes to global warming. ✓

(d) Allow C_xH_y to be the unknown hydrocarbon product:

$$C_{10}H_{22} \rightarrow C_5H_{10} + C_xH_y$$

Remember that the equation must balance so the number of the carbons and hydrogens on both sides must equal.

$10 = 5 + x$ $\qquad$ $22 = 10 + y$

$x = 5$ $\qquad\qquad$ $y = 12$

$$C_{10}H_{22} \rightarrow C_5H_{10} + C_5H_{12} ✓$$

Question 2

C_2H_5SH, a compound with an unpleasant smell, is added to gas to enable leaks from gas pipes to be more easily detected.

(a) Write an equation for the combustion of C_2H_5SH. (1 mark)

(b) Identify a compound among the combustion products that is toxic and leads to pollution. Explain how it can be removed. (3 marks)

ⓔ In this question an organic compound, containing sulfur, is burnt. You must realise that sulfur dioxide is produced in addition to carbon dioxide and water. Write the formula in the equation, and then remember to balance. You can use half numbers to balance. The word 'identify' in the question means that you can name the compound or give its correct formula. The toxic gas is sulfur dioxide, which is acidic so it is removed by neutralisation with a base.

(a) $C_2H_5SH + 4\frac{1}{2}O_2 \rightarrow 2CO_2 + 3H_2O + SO_2$ ✓

(b) Sulfur dioxide ✓; neutralisation ✓; with calcium oxide or calcium carbonate ✓.

Question 3

Which one of the following can act as a free radical? (1 mark)

A Cl

B Cl⁺

C Cl⁻

D Cl_2

The correct answer is A ✓

ⓔ A radical is a species with an unpaired electron. You need to work out the number of electrons in the outer shell of each species; Cl has seven outer shell electrons, so it has one unpaired electron and is a radical. Cl⁺ has lost one electron so it has six electrons in the outer shell — they are all paired so it is not a radical. Cl⁻ has gained an electron so it has eight in the outer shell and four pairs, so it is not a radical. In Cl_2 both chlorine atoms share electrons, so there are eight in the outer shell — again no unpaired electrons.

Halogenoalkanes

Question 1

The flow scheme in Figure 4 shows some of the reactions of 1-bromobutane. Draw the structure of the organic products from the reactions shown. (4 marks)

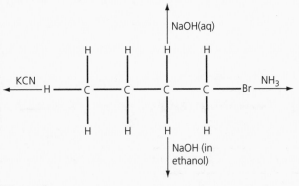

Figure 4

ⓔ This type of question, involving different reactions of a particular homologous series, is common. Most reactions of halogenoalkanes are substitution reactions, in which the halogen atom is substituted by a different group (OH, NH_2 and CN in these examples). With hydroxide ions dissolved in ethanol, an elimination reaction occurs in which the hydrogen halide (HBr in this case) is eliminated to give an alkene.

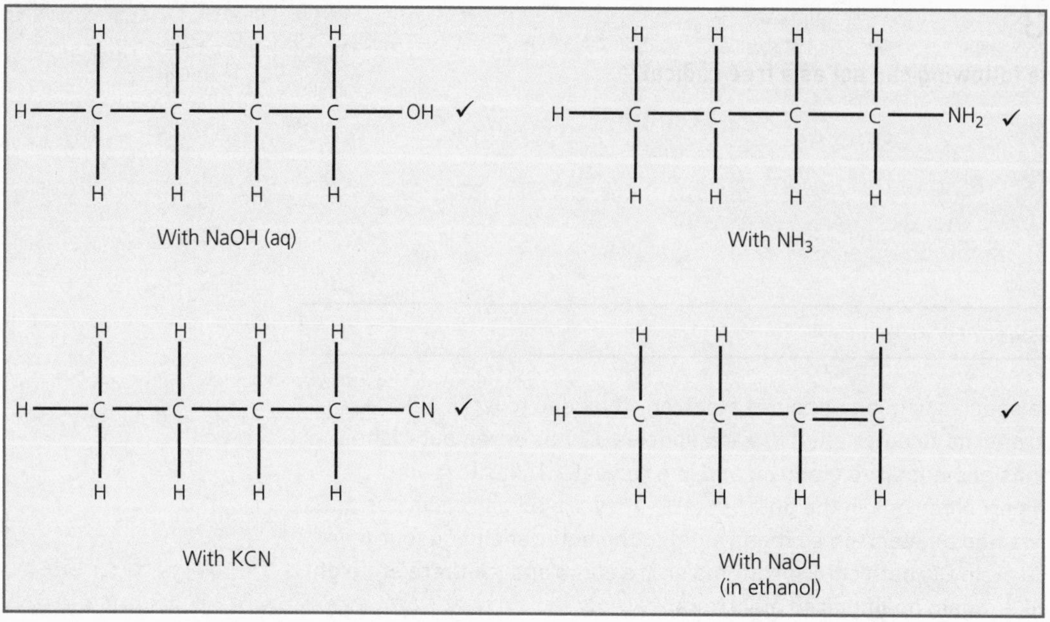

With NaOH (aq)

With NH$_3$

With KCN

With NaOH (in ethanol)

Question 2

Which of the following describes the mechanism for the reaction between aqueous hydroxide ions and 1-bromopropane?

(1 mark)

A   electrophilic addition

B   electrophilic substitution

C   nucleophilic addition

D   nucleophilic substitution

ⓔ Halogenoalkanes react with aqueous sodium hydroxide by nucleophilic substitution. It is important that you can recognise the compound as a halogenoalkane and relate its reaction with aqueous hydroxide ions to the type of mechanism.

Answer is D ✓

Question 3

Outline a mechanism for the reaction in which 2-bromopropane (CH$_3$CHBrCH$_3$) reacts with ethanolic potassium hydroxide.

(3 marks)

ⓔ It is important that you realise that the reaction is with ethanolic potassium hydroxide, so it is an elimination reaction, producing propene. To outline a mechanism it is sufficient to give the diagram shown in the student's answer, making sure that the curly arrows come from a bond or a lone pair.

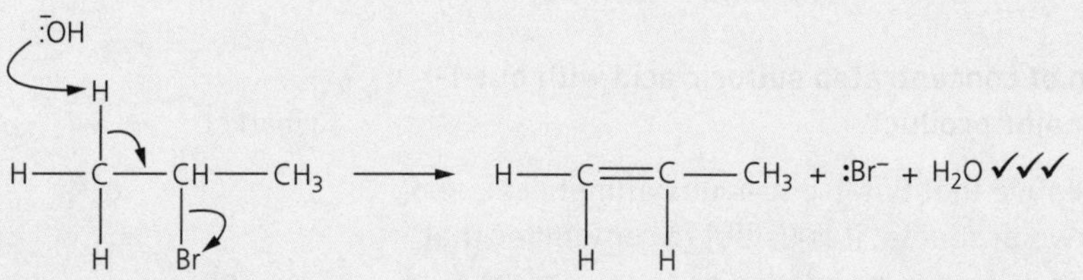

Question 4

Equal volumes of 1-chlorobutane and 1-iodobutane are warmed with aqueous silver nitrate in the presence of ethanol. Which one of the following is the reason why the 1-chlorobutane reacts more slowly?

(1 mark)

A The C–Cl bond is more polar than the C–I bond.

B The C–Cl bond is stronger than the C–I bond.

C The C–I bond is more polar than the C–Cl bond.

D The C–I bond is stronger than the C–Cl bond.

The correct answer is B ✓

ⓔ 1-chlorobutane and 1-iodobutane are halogenoalkanes and are hydrolysed by the aqueous solution of silver nitrate — a white precipitate will be produced slowly for 1-chlorobutane, and a yellow precipitate will be produced quickly for 1- iodobutane. This is because the bond enthalpy of the C–I bond in 1-iodobutane is less than that of the C–Cl bond in 1-chlorobutane — in other words the C–Cl is stronger. It is not explained in terms of polarity.

Alkenes

Question 1

Which of the following describes the mechanism of the reaction between ethene and hydrogen bromide?

(1 mark)

A electrophilic addition

B electrophilic substitution

C nucleophilic addition

D nucleophilic substitution

ⓔ You need to recognise that all the reactions of alkenes that you have studied are addition reactions producing a single product. This eliminates answers C and D. An electrophile is an electron donor and attacks a centre of negative charge, such as a double bond.

The answer is A ✓

Question 2

Outline the mechanism for the reaction of concentrated sulfuric acid with but-1-ene, to illustrate the formation of the major product.

(3 marks)

e To answer this question you must realise that but-1-ene is unsymmetrical, and the electrophilic addition will produce two products. It is useful to remember that the major product is the one in which the hydrogen has added on to the carbon with the most hydrogens that are directly attached. The major product is produced via the most stable carbocation, that is, the secondary carbocation. Therefore, in the mechanism it is the secondary carbocation that you must include. Be careful when drawing curly arrows. They must come from the centre of a bond or from a lone pair of electrons.

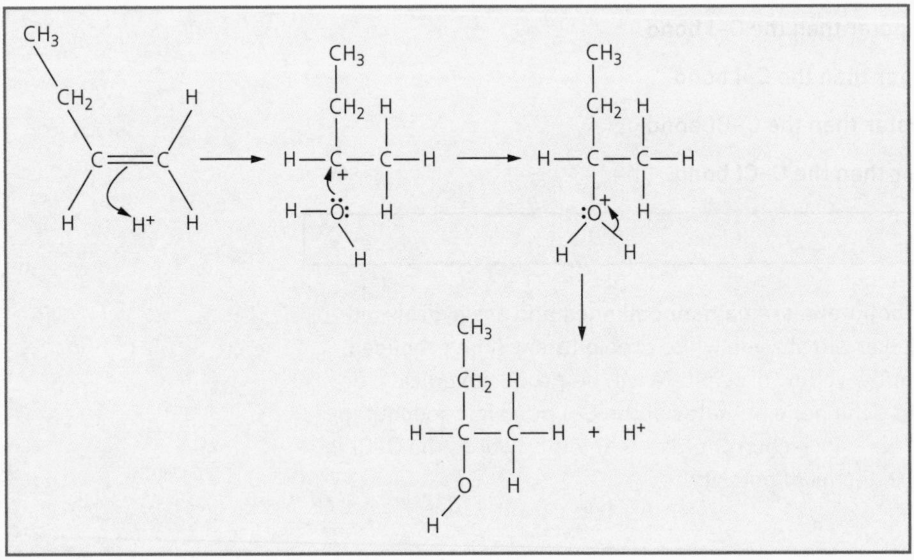

Question 3

The alkene *E*-but-2-enenitrile (Figure 5) is used to make acrylic plastics.

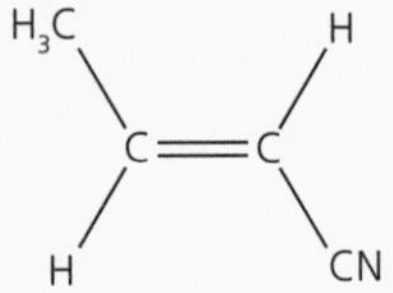

Figure 5

(a) (i) Draw the structure of *Z*-but-2-enenitrile. (1 mark)

(ii) Identify the feature of the double bond in the *E* and *Z* isomers that causes them to be stereoisomers. (1 mark)

(b) Draw the repeating unit of the poly(alkene) formed by addition polymerisation of *E*-but-2-enenitrile. (1 mark)

ⓔ To draw *E–Z* isomers you need to identify the highest priority group on each carbon of the double bond. The highest priority group is the one with the largest atomic number. Consider the left-hand carbon — attached to it is a C (atomic number 6) and H (atomic number 1). The C of the CH_3 has highest priority. Consider the right-hand carbon — attached is a H (atomic number 1) and C (atomic number 6). The C of the CN has highest priority. The *Z* isomer has the highest priority groups on the same side of the double bond.

(a) (i) The structure is:

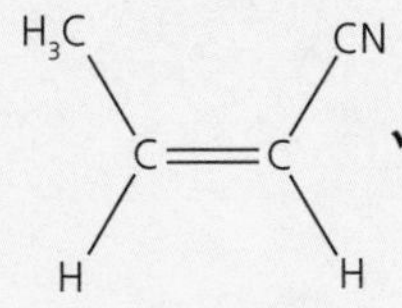

(ii) There is restricted rotation/no rotation about the double bond ✓

(b)

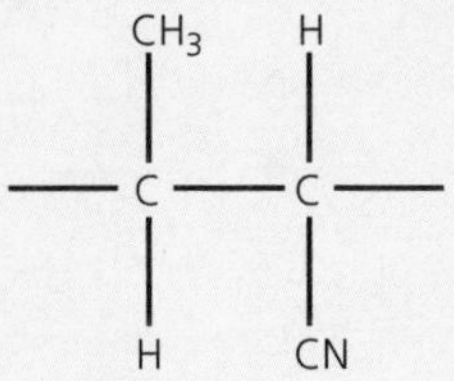

Exam tip

To draw the repeating unit simply remove the double bond and place bonds on either side of the carbons. Remember, you will be penalised if you include an 'n' — as you are only asked for the repeating unit, not the polymer.

Question 4

The reactions of methane and ethene are different. Methane reacts with chlorine in the presence of ultraviolet light, and ethene reacts readily with hydrogen bromide.

(a) (i) **Name the mechanism by which methane reacts with chlorine in the presence of ultraviolet light.** (2 marks)

(ii) **Explain this mechanism.** (6 marks)

ⓔ Comparing the reactivity of alkanes and alkenes is a common question. Alkanes are much less reactive than alkenes and undergo substitution reactions; the C=C in alkenes allows them to undergo addition reactions. Part (a) requires the mechanism for an alkane.

(a) (i) Free-radical ✓ substitution ✓

e Remember, in a mechanism for alkanes it is sufficient to name each step — initiation, propagation and termination — and to state equations for each step. No words of explanation are necessary. You must show a • on each radical, but it can be in any position.

> **(ii)** Initiation ✓, $Cl_2 \rightarrow 2Cl•$ ✓
> Propagation ✓, $CH_4 + Cl• \rightarrow CH_3• + HCl$ ✓
> $CH_3• + Cl_2 \rightarrow Cl• + CH_3Cl$ ✓
> Termination, $CH_3• + Cl• \rightarrow CH_3Cl$
> $CH_3• + CH_3• \rightarrow C_2H_6$
> $Cl• + Cl• \rightarrow Cl_2$ (any one termination step) ✓

(b) (i) Write an equation for ethene reacting with chlorine, and name the organic product. (2 marks)

(ii) Name the mechanism for the reaction between ethene and chlorine. (2 marks)

e When ethene reacts with chlorine an addition reaction occurs. The chlorine adds on across the double bond. There are only two carbons so one chlorine bonds to each carbon. It is best to give structural formulae in your answer.

> **(b) (i)** $CH_2=CH_2 + Cl_2 \rightarrow CH_2ClCH_2Cl$ ✓; 1,2-dichloroethane ✓
> **(ii)** Electrophilic ✓ addition ✓

Alcohols

Question 1

Which one of the following is formed by the oxidation of propan-2-ol using acidified potassium dichromate(VI) solution when heated under reflux? (1 mark)

A propanoic acid

B propanone

C propanal

D ethanoic acid

> The answer is B ✓

e It is useful to draw out the structure of propan-2-ol, as shown in Figure 6.

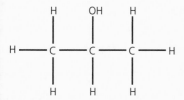

Figure 6

It is then easy to see that it is a secondary alcohol — it has two alkyl groups bonded to the carbon that is bonded to the OH group. Secondary alcohols are oxidised to ketones.

Question 2

Ethanol can be oxidised by acidified potassium dichromate(VI) to either ethanal or ethanoic acid.

State and explain the reaction conditions necessary for oxidation to ethanal and to ethanoic acid. Give equations for both oxidations using [O] to represent the oxidising agent.
(6 marks)

ⓔ In this longer answer question you are expected to do several things: state, explain, and give equations. It may be best to highlight or underline the key words in the question. You need to realise that ethanol is oxidised via the aldehyde, ethanal, to the acid, thus removing it from contact with the oxidising agent as soon as it is formed by distillation will prevent it from being oxidised further to the acid. Reflux is necessary to oxidise ethanol completely to ethanoic acid. The repeated boiling of the ethanol and oxidising agent and condensation of the vapour ensures that the reactants are in prolonged contact to cause complete oxidation, and any ethanal and ethanol that initially evaporate can be condensed and oxidised.

- Oxidation to ethanal — heat with oxidising agent in distillation apparatus ✓, which removes the aldehyde from the oxidising mixture immediately ✓.

- Oxidation to ethanoic acid — use excess oxidising agent and heat under reflux (repeated boiling and condensing of a reaction mixture) ✓, which ensures complete oxidation, and that any ethanal and ethanol that initially evaporate can be condensed and oxidised ✓.

$$CH_3CH_2OH + [O] \rightarrow CH_3CHO + H_2O ✓$$
$$CH_3CH_2OH + 2[O] \rightarrow CH_3COOH + H_2O ✓$$

Question 3

1-bromobutane can be prepared by heating a mixture of butan-1-ol and hydrogen bromide under reflux. The impure 1-bromobutane is distilled off and washed with sodium carbonate solution before being mixed with anhydrous sodium sulfate.

(a) Write an equation for the preparation of 1-bromobutane. (1 mark)

(b) What is meant by 'heating under reflux'? (1 mark)

(c) The sodium carbonate solution is mixed with the impure organic compound in a separating funnel.

   (i) Explain why the mixture is washed with sodium carbonate solution.   (1 mark)

   (ii) After washing, the mixture separates into two layers. Explain how you would identify the organic layer.   (2 marks)

(d) What is the purpose of the anhydrous sodium sulfate?   (1 mark)

ⓔ Questions on experiments are common. You need to understand the purpose of each process. The actual experiment may not be the one that you carried out in school, but the processes within it to produce a pure organic liquid (or solid) are the same. Purification is carried out in a separating funnel, often by shaking with an aqueous solution — and separating the layers. Anhydrous sodium sulfate is a common drying agent, which is used to remove water from an organic liquid. This reaction is not on your specification, however, simply use the information in the question. You know the formula of butan-1-ol, $CH_3CH_2CH_2CH_2OH$, and of 1-bromobutane, $CH_3CH_2CH_2CH_2Br$ — put these into an equation.

(a) $CH_3CH_2CH_2CH_2OH + HBr \rightarrow CH_3CH_2CH_2CH_2Br + H_2O$ ✓

(b) Repeated boiling and condensing of a reaction mixture ✓

ⓔ You must realise that sodium carbonate solution is a base, so it reacts with acidic impurities when mixed with the organic compound.

(c) **(i)** To remove acid ✓

   (ii) Add water; the layer that increases in volume is the aqueous layer. ✓ The other layer is organic. ✓

(d) To remove water/drying agent ✓

Question 4

Ethanol is formed by the fermentation of glucose.

(a) Write an equation for the reaction occurring during fermentation.   (1 mark)

(b) In industry, this fermentation reaction is carried out at 35°C rather than at 25°C. Suggest one advantage and one disadvantage for industry of carrying out the fermentation at this higher temperature.   (2 marks)

(c) Suggest why the ethanol produced would be contaminated by ethanoic acid.   (1 mark)

ⓔ The equation for fermentation is one which must be known. In terms of the temperature used, a higher temperature always produces a higher rate, but at the expense of using more energy. The ethanol produced in fermentation is open to oxidation by the air to form ethanoic acid.

(a) $C_6H_{12}O_6(aq) \rightarrow 2CH_3CH_2OH(aq) + 2CO_2(g)$ ✓

(b) Advantage — ethanol is produced at a faster rate. ✓

Disadvantage — more energy is used/required in the reaction. ✓

(c) It is oxidised by the air. ✓

Organic analysis

Question 1

Describe how a student could use chemical tests to confirm that a liquid was propanal and not propanoic acid. (5 marks)

ⓔ First, it is best to decide what homologous series the liquids belong to — propanal is an aldehyde (-al) and propanoic acid a carboxylic acid. A common error is to outline a positive test for propanal, without giving a test to prove it is not propanoic acid. Both are needed. You must be familiar with the table of organic tests on page 61. To test for an aldehyde it is possible to use either Tollens' reagent or Fehling's solution, and both must be warmed with the test sample. To test for an acid, adding a carbonate gives bubbles of carbon dioxide in a positive test.

Add Tollens' reagent ✓; silver mirror ✓.

or

Add Fehling's solution ✓; red precipitate ✓. In either case, the reagent and the sample must be warmed. ✓

To confirm the absence of ethanoic acid, add sodium hydrogen carbonate or sodium carbonate. ✓ No effervescence observed. ✓

Question 2

(a) Draw the structure of the alkene that has a peak, due to its molecular ion, at $m/z = 42$ in its mass spectrum. (1 mark)

ⓔ An alkene has a general formula C_nH_{2n}, and its m/z value gives the M_r. C_2H_4 has $M_r = 28$, but C_3H_6 has $M_r = 42$. To answer the question you must draw the structure of C_3H_6 showing the double bond. Make sure the double bond is between the carbons.

(a) Structure for propene is $H_2C=CHCH_3$ ✓

(b) Draw the structure of the organic product with $M_r = 73$, which is made from the reaction between 2-bromobutane and ammonia. (1 mark)

ⓔ You need to be familiar with the nucleophilic substitution reactions of halogenoalkanes and ammonia to produce an amine. The NH_2 group substitutes the bromine. It is helpful to write an equation and then check if the M_r of the product is 73.

$$CH_3CH_2CHBrCH_3 + 2NH_3 \rightarrow CH_3CH_2CHNH_2CH_3 + NH_4Br$$

(b) $CH_3CH_2CHNH_2CH_3$ ✓

(c) Draw the displayed formula of the organic product formed when propan-1-ol is completely oxidised. Give two differences between the infrared spectrum of propan-1-ol and this product, other than in their fingerprint regions. (3 marks)

ⓔ Propan-1-ol is a primary alcohol and it can be completely oxidised to give propanoic acid. Remember that in a displayed formula every bond must be shown. To state the differences between the infrared spectra of the alcohol and of the acid, make sure that you use your data sheet to quote values. For the acid there will be a C=O and O–H absorption. The OH absorption for alcohols will be in a different place.

(c)

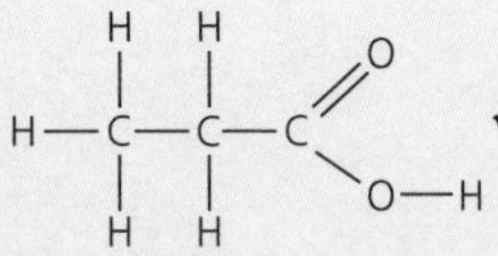

Alcohol OH absorption in different place $(3230–3550\,cm^{-1})$ from that of the acid OH absorption $(2500–3000\,cm^{-1})$ ✓

The C=O in acids has an absorption at $1680–1750\,cm^{-1}$, which is not found in alcohols. ✓

Question 3

Which one of the following occurs when a molecule absorbs infrared radiation? (1 mark)

A Electrons in the bonds are excited.

B The bonds bend and eventually break.

C The bonds rotate.

D The bonds vibrate.

The correct answer is D ✓

ⓔ This is a factual question — you must know that a molecule can absorb infrared radiation because the bond can vibrate.

Knowledge check answers

1 Period 3 and p block

2 Calcium has a smaller ionic radius than strontium and the delocalised electrons are closer to the positive nucleus, the metallic bond is therefore stronger and needs more energy to break.

3 $Sr + 2H_2O \rightarrow Sr(OH)_2 + H_2$

4 $Mg(OH)_2$

5 It will react with the barium ions and produce a white precipitate of barium sulfate, ruining the test.

6 Fluorine

7 Chlorine gains electrons and is reduced/it is a reduction reaction.

8 $KBr + H_2SO_4 \rightarrow KHSO_4 + HBr$

9 Yellow precipitate, which remains on addition of dilute ammonia solution.

10 Sodium chlorate(I), sodium chloride, water

11 $C_{10}H_{22}$

12 Molecular formula: $C_4H_7Br_2Cl$
Empirical formula: $C_4H_7Br_2Cl$
Structural formula: $CH_2CClBrCH_2CH_2Br$

13 Cl

14 2,2-dibromobutane

15 a 2-chloropropan-1-ol
b 2-methylpropanal

16 Butanoic acid; 2-methylpropanoic acid

17 a But-2-ene; b 2-methylpropene

18 Hex-2-ene; hex-3-ene

19 Restricted rotation about the planar carbon–carbon double bond.

20 C_3H_8; because it has a lower boiling point as it has lower M_r and weaker van der Waals forces between its molecules.

21 $C_8H_{18} \rightarrow C_2H_4 + C_3H_6 + C_3H_8$ (propane)

22 $C_8H_{18} + 25NO \rightarrow 8CO_2 + 12\frac{1}{2}N_2 + 9H_2O$; the catalyst is platinum or rhodium.

23 Sulfur dioxide. Neutralise by passing through a slurry of calcium oxide or calcium carbonate.

24 A species with an unpaired electron; Br•

25 The halogen is more electronegative than carbon and draws the electrons in the bond to itself, creating a small negative charge and leaving a small positive charge on the carbon.

26 1-bromobutane; the C–Br bond has a lower bond enthalpy than the C–Cl bond.

27 $CH_3CH_2CH_2Br + 2NH_3 \rightarrow CH_3CH_2CH_2NH_2 + NH_4Br$; propylamine

28 $CH_3CH_2CH_2CH_2Br + KOH \rightarrow C_4H_8 + KBr + H_2O$

29 a Butene/but-1-ene/but-2-ene
b Butan-1-ol

30 The carbon–chlorine bonds in CFCs are broken by ultraviolet light, producing chlorine radicals that catalyse the decomposition of ozone and contribute to holes in the ozone layer.

31 3-chlorobut-1-ene; it contains chlorine, carbon and hydrogen atoms, not carbon and hydrogen only.

32 Electrophilic addition

33 Propyl hydrogensulfate

34 Bromine water changes from orange to colourless in cyclohexene but no change in cyclohexane.

35 $H_2C=CHCH(CH_3)_2$

36 An addition polymer is a long-chain molecule formed by repeated addition reaction of many alkene monomers.

37 Yeast, no air, 35°C

38 $C_4H_9OH + 6\frac{1}{2}O_2 \rightarrow 4CO_2 + 5H_2O$
An excess of oxygen
It can be made from plant material

39 The position of equilibrium will move from left to right to remove the excess steam and give a higher yield of product.

40 Butanol can form hydrogen bonds with water owing to its polar OH group. Butane is non-polar and cannot form hydrogen bonds with water.

41 $CH_3CH_2CH_2CH_2OH + 2[O] \rightarrow CH_3CH_2CH_2COOH + H_2O$

42 Warm with Tollens' reagent — the solution should remain colourless. Or, warm with Fehling's solution — the solution should remain blue.

43 Warm with acidified potassium dichromate(VI) solution
For 2-methylpropan-2-ol (tertiary alcohol) it should remain an orange solution
For the ethanol (primary alcohol) it should change colour from an orange solution to a green solution.

44 60.0

45 A sharp, intense C=O peak at 1680–1750 cm^{-1}

Index

Index